18th Topical Conference on Gas Utilization 2018

Topical Conference at the 2018 AIChE Spring Meeting and 14th Global Congress on Process Safety

Orlando, Florida, USA
22 - 25 April 2018

ISBN: 978-1-5108-6426-9

Printed from e-media with permission by:

Curran Associates, Inc.
57 Morehouse Lane
Red Hook, NY 12571

Some format issues inherent in the e-media version may also appear in this print version.

Printed by Curran Associates, Inc. (2018)

For permission requests, please contact AIChE
at the address below.

AIChE
120 Wall Street, FL 23
New York, NY 10005-4020

Phone: (800) 242-4363
Fax: (203) 775-5177

www.aiche.org

Additional copies of this publication are available from:

Curran Associates, Inc.
57 Morehouse Lane
Red Hook, NY 12571 USA
Phone: 845-758-0400
Fax: 845-758-2633
Email: curran@proceedings.com
Web: www.proceedings.com

TABLE OF CONTENTS

Some abstracts in this book refer to images that were not made available by the publisher

513650 Adsorbents for Trace Sulfur Removal for Gas Processing

Abstract:

Honeywell UOP provides a full range of solutions for natural gas conditioning including the use of adsorbents for trace contaminant removal. Trace contaminants in natural gas can cause a range of issues including corrosion, environmental challenges, and the inability to meet product specifications. Sulfur compounds are one of the noteworthy contaminants present in natural gas. A variety of sulfur compounds are pervasive in natural gas and natural gas liquids including hydrogen sulfide, carbonyl sulfide, mercaptans, organic sulfides and disulfides. A range of process options are needed to meet product sulfur specifications since no single solution will work for removal of all of the sulfur compounds in the wide range of contaminant concentrations and hydrocarbon compositions found in natural gas globally. In order to meet these needs, UOP has developed a range of processing options, including use of membranes, solvents, and adsorbents to produce natural gas to meet product quality specifications for saleable natural gas and natural gas liquid products.

Discussion will focus on processing options for removal of trace sulfur compounds from natural gas and natural gas liquids. Regenerable and non-regenerable adsorbent alternatives are capable of achieving tight specifications cost effectively. Regenerable molecluar sieve adsorbents are particularly effective for mercaptan removal even when present at hundreds of ppm in some regions of the world. UOPâ€™s improved adsorbent design configurations can mitigate contaminant spikes upon thermal regeneration. Non-regenerative approaches to trace H_2S and mercaptan removal can allow meeting copper strip specifications for natural gas liquids.

511173 Tailoring Amine Based Solvents for Reactive CO_2 Absorption Using Structure-Property Relationships

Abstract:

The development of new amines for CO_2 absorption for post combustion carbon capture processes to mitigate the climate change has received much attention in the last years. But CO_2 absorption also remains an important task in many established large scale industrial processes like the purification of synthesis gas. Such processes could benefit from tailored new amines.

In the present project novel amines for the removal of CO_2 from synthesis gas by reactive absorption are developed and tested. All studied amines are derivatives of triacetone amine (TAA). These interesting amines have not been considered before as reactive solvents for CO_2 capture. The approach of the present work is new: we have realized a shake-hand between the knowledge-guided synthesis of the new TAA-based amines and their characterization as solvents.

In a first step, a short list of TAA-based amines, that looked promising, was established, and the possibilities to synthetize them were evaluated. Some of the candidates were then synthesized and studied experimentally. Based on the results of these studies, new candidates were identified. In the iterative process, about 50 amines were considered and about 25 of them were actually synthesized and investigated. These amines differ only in the substituent of the basic TAA ring structure. From the results relationships between the structure of the amines and their properties as reactive solvents for CO_2 capture can be established. Some of these structure-property relationships can be generalized.

The experimental methods used for the characterization of the solvents have been described in detail recently [1]. The measurements include: solubility and rate of absorption of CO_2, liquid-liquid and solid-liquid equilibrium and viscosity. Furthermore, basic physico-chemical data like acid constants, and chemical equilibrium constants, namely those of the carbamate formation, were determined. The same procedures were applied for all amines.

Based on this comprehensive data the amines were assessed regarding their performance as solvents. An enhanced version of the method of Notz et al. [2] which allows ranking solvents with respect to the process energy requirement and solvent flow rate was employed. Some of the particularly attractive solvents show a temperature-dependent liquid-liquid phase split, which could be used beneficially in the reactive absorption process. The most promising solvent was successfully tested in pilot scale.

[1] D. Vasiliu, A. Yazdani, N. McCann, M. Irfan, R. Schneider, J. Rolker, G. Maurer, E. von Harbou, H. Hasse: Thermodynamic Study of a Complex System for Carbon Capture: Butyltriacetonediamine + Water + Carbon Dioxide, J. Chem. Eng. Data 61 (2016) 3814â€“3826

[2] R. Notz, I. TÃ¶nnies, H. Mangalapally, S. Hoch, H. Hasse: A short-cut method for assessing absorbents for post-combustion carbon dioxide capture, Greenhouse Gas Control 5 (2011) 413-421

513594 Adsorptive Removal of CO_2 from Natural Gas

Abstract:

Raw natural gas contains methane as its major component, but it also contains considerable amounts of contaminants such as CO_2 and H_2S (i.e. acid gases) that can cause corrosion and fouling of the pipeline and equipment during transportation and liquefaction. Amine-based CO_2 gas removal processes have been employed in the gas industry, but these processes have disadvantages including high regeneration energy requirements and inefficiencies; these issues have not been adequately solved to date. Currently, adsorptive acid gas removal technologies have received significant interest because of the simplicity of adsorbent regeneration by thermal or pressure variation[1]. Numerous micro- and mesoporous adsorbents including zeolites [2-3], titanosilicates[4], activated carbons[5-6], metal-organic-framework (MOF) [7], and silica-alumina materials[8-9] were studied for this type of application. However, the CO_2/CH_4 selectivity of the aforementioned adsorbents was not high enough for commercial applications. In this study, different adsorbents including cation-exchanged zeolites and amine-modified MOFs were synthesized, physicochemically characterized, and evaluated for adsorptive removal of CO_2 from natural gas. The equilibrium isotherms of CO_2 and CH_4 on adsorbents in varied pressure were measured and compared. The ideal-adsorbed solution theory (IAST) was employed for the estimation of CO_2/CH_4 selectivity of those adsorbents.

References:

1) D. Aaron, C. Tsouris, (2005) *Separ. Sci. Technol.* , 40, 321–348

2) J. Collins, US Patent No. 3,751,878. 1973.

3) M. W. Seery, US Patent No. 5,938,819. 1999

4) W. B. Dolan, M.J. Mitariten, US Patent No. 6,610,124 B1. 2003

5) A. Kapoor, R.T. Yang, *Chem. Eng. Sci.* 1989, 44, 1723–1733

6) A. Jayaraman, Chiao, A. S.; Padin, J.; Yang, R. T.; Munson, C. L., *Separ. Sci. Technol.* (2002), 37, 2505–2528

7) L. Hamon, E. Jolimaitre, G. Pringruber , (2010), *Ind. Eng. Chem. Res.*, 49, 7497-7503

8) W.B. Dolan, M.J. Mitariten, US patent No. 2003/0047071, 2003

9) G. Bellussi, P. Broccia, A. Carati, R. Millini, P. Pollesel, C. Rizzo, M. Tagliabue, (2011), Micropor. Mesopor. Mat., 146, 134–140

MEETING ADDITIONAL PLANT CAPACITY REQUIREMENTS WITHOUT ADDING ADDITIONAL VESSELS

Gene Eberhardt
Woven Metal Products - Alvin, TX USA

Gene Eberhardt is the Vice President of the Reactor Division for Woven Metal Products in Alvin, Texas. He holds a Bachelor of Science Degree in Business from Saint Mary's University of Minnesota, a Master's Degree in Business Administration (MBA) from Purdue University and a Master's Degree in International Business Management from TiasNimbas Business School - The Netherlands. He holds patents for Catalyst Bed Support systems, Systems and Methods for Improving Flow in Radial Flow Reactors and Scallop Support Distributor for Radial Flow Reactors. He continues to develop industry changing technologies and methods in collaboration with Purdue University.

Given today's demanding throughput needs to maximize plant profits, plant de-bottlenecking quickly becomes an area ideal for identification and immediate return on investments. Today's gas plants require systems optimization to safely produce and deliver predictable, profitable products, longer predictable drying cycles and increased molecular sieve bed performance. These are not off-the-shelf solutions and require robust engineered solutions that are performance driven, industry proven and readily available for demanding plant outage requirements.

Molecular sieve bed support is one of, if not the most critical mechanical components within the dehydration vessel. Once the bed support requirements are understood, all options can be evaluated and a competent molecular sieve bed plan can be developed.

As with many natural gas plants commissioned over the years, molecular sieve bed support and proper molecular sieve bed utilization have been a secondary concern in these plant designs. In many cases, operational conditions have not allowed for these units to be properly evaluated or engineered to meet the operational needs of the plant. It has become an industry standard to install an outlet collector (elephant stool) in the outlet nozzle of the vessel with very little engineering emphasis given to the outlet collector, the effects that the outlet collector has on the molecular sieve bed, the performance of the molecular sieve bed or the effects the outlet collector has on the life of the molecular sieve bed. In many applications, the outlet collector serves only as a containment device and has not been engineered to aid in the bed performance or in the bed longevity. Many of the outlet collectors utilized today include an assortment of materials and configurations that include (but not limited to) slotted pipe, rolled plate, engineered support frames and a variety of material coverings including – wire mesh in a variety of material grades, wire mesh sizes and open area. In addition to the many wire mesh offerings many licensors and fabricators select profile wire in a wide range of stainless steel material grades and slot openings all of which do not observe any industry standards or a standard engineering design for these applications.

In some applications, a dehydration unit will have a horizontal bed support installed near the tangent line or the head seam of the vessel with a variety of materials used for bed containment including layers of wire mesh, wire tied over grating and profile wire grid arrangements. In these bed configurations, the outlet collector is often installed as a secondary containment device as most vessels are not rolled perfectly round creating challenges to accomplish a full perimeter seal. The horizontal bed support configuration is a significant upgrade to the traditional outlet collector only configuration, however the molecular sieve bed design and perimeter sealing must be fully understood with emphasis on a secondary containment device to prevent total bed migration if the perimeter seal becomes compromised. The horizontal bed support option does offer much better bed utilization; however, this does sacrifice the total bed volume as the head of the vessel (at a minimum) becomes a void space.

This paper will focus on a recent installation of the Woven Metal Products / Haldor Topsoe co-developed (U.S. Patent Number 9409137) Outlet Support Grid (OSG) bed support system for a Midstream Company in the North East United States. This complex, which consists of two cryogenic natural gas processing plants, each of which has 200 MMcf/d of processing capacity. This plant design includes 3 industry standard dehydration vessels that have an inner diameter of 84 inches (7 feet) and a total seam length (not including vessel heads) of 164 inches (13 feet 8 inches). When operational, this plant has been designed for parallel adsorption (Down Flow) and one vessel in regeneration (Up Flow) / standby.

Woven Metal Products Outlet Support Grid U.S. Patent 09409137

The Outlet Support Grid now offers an engineered total bed solution utilizing the performance of a horizontal bed support with the containment reliability that an outlet collector provides. Often the additional bed volumes realized with the Outlet Support Grid system installed will provide 15% or more additional bed volumes depending on the elevation of the horizontal bed configuration it replaces. As shown in Figure 1 the Computational Fluid Dynamics (CFD) modeling of the Outlet Support Grid shows a very uniform flow pattern which is critical for optimized bed performance, combined with full bed containment the Outlet Support Grid system is a reliable operational solution.

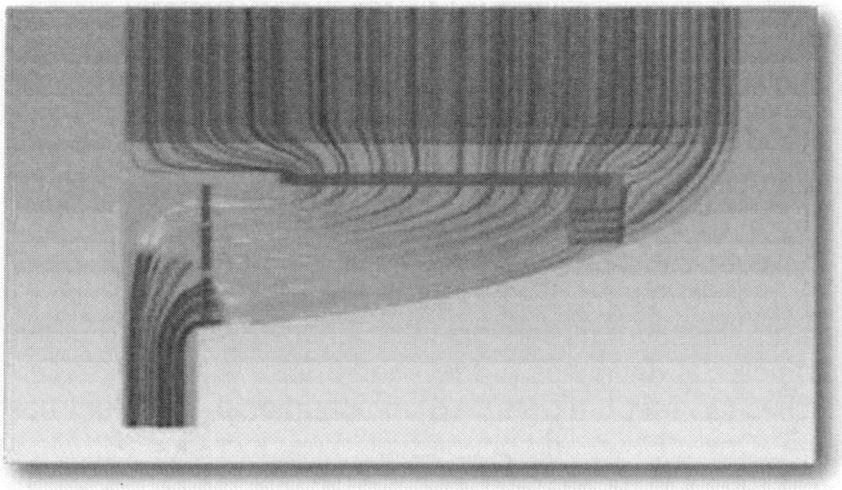

Figure 1.

History

This natural gas plant design was commissioned in April 2015, at which time operations had encountered / suspected a bed lifting (bed "fluffing") that resulted in an unusually high pressure reading across all three molecular sieve beds of 130 psi ΔP indicated. This unusually high ΔP was also accompanied with heavy dusting and plugging in the units down-stream filters, which prompted the operations team to suspect molecular sieve crushing to have incurred during the startup of the plant.

New molecular sieve was promptly ordered for the three dehydration vessels and the unit outage was taken for the replacement of the suspected crushed molecular sieve. During this outage, the unit beds were not fully emptied or cleaned only the molecular sieve was replaced and then the beds were put back in service with the existing bed support media undisturbed and no inspections or cleaning of the outlet baskets or inert media was performed. These three units were restarted and the high ΔP remained at 130 psi and continued to climb.

The operations team was now faced with the task of discovering any and all options to correct the high ΔP situation and restore the plant to a safe operational rate, quickly. The operations and engineering teams contacted their local UNIVAR field representative to source not only the new molecular sieve and new inert media support balls for each dehydration bed but they also started looking for replacement options for the potential of a semi crushed or fully crushed outlet basket in each of the three dehydration units. Woven Metal Products was contacted and was informed of the current operational situation and the challenges facing the operations team.

Collectively the Midstream company, Woven Metal Products, UNIVAR, Zeochem and Catalyst Force teams addressed the following items; safety risks of continued operations, critical project timeline, operational limitations, current bed configuration, and equipment history of the installed equipment.

Each of the three dehydration vessels were originally equipped with outlet baskets sized at 15 inches outer diameter, 10 inches overall height with a profile wire covering, then welded to the outlet nozzle. These outlet baskets were constructed of 304 stainless steel, designed for 50 PSIG and 30% - 40% open area. Each dehydration vessel was loaded as follows:

Pre – Installation Bed Loading			
Media	**Pounds**	**Cubic Feet^3**	**Feet / Inches**
Ceramic Inert Balls (size unknown)	820 (estimate)	19	0.5 Feet
Guard Bed Media	1,653	37.57	0.98 Feet
4 x 8 mesh 1/8" Molecular Sieve	12,600	273.91	7.12 Feet
8 x 12 mesh 1/16" Molecular Sieve	7,200	153.19	3.98 Feet
Ceramic Inert Balls (size unknown)	820 (estimate)	19	0.5 Feet
Totals	23,093 (estimate)	502.67	13.08 Feet

The operational records reveal an average ΔP of 134 PSIG with a maximum ΔP of 180 PSIG indicated in table 1 (below) given the very high ΔP the actual condition of the outlet basket and molecular sieve bed was unknown. The operations reported a high frequency of heavy filter dusting / plugging, exchangers plugging, high system pressure requirements, dew point control issues, overall output product controls and longer bed regeneration times.

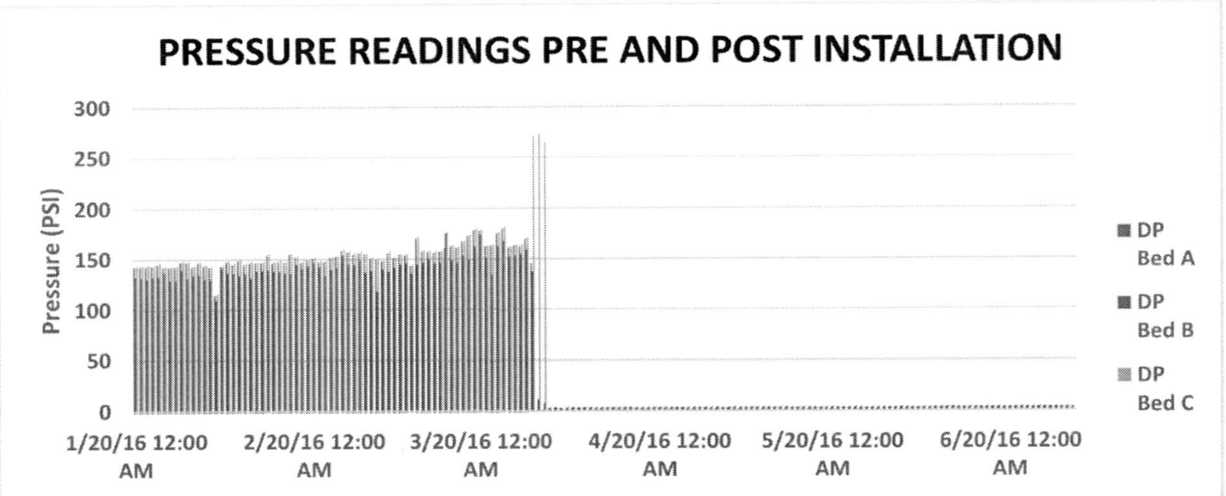

Table 1.

The following options were considered given the current vessel configuration and the critical path for both fabrication and the unit outage requirements;

- Replace the current outlet baskets with a design that would allow more open area and sized to the vessel.
- A horizontal support grid option was not considered due to the dehydration vessels not originally equipped with a support ring or beams and the time required to modify the vessels.
- Outlet Support Grid (OSG) system.

Given the Midstream companies operational experience with the outlet baskets the evaluation and decision was made to install the Outlet Support Grid solution. Woven Metal Products began engineering the Outlet Support Grid solution with fabrication drawings provide to the client for approvals in two weeks and full fabrication / delivery promised in just four weeks. The team confirmed that the Zeochem molecular sieve and bed support media would be onsite for the outage date and the Catalyst Force team would be able to service the outage with the required staff to execute a safe and timely outage. Schedules were drafted for the expedited fabrication and for the unit outage, both with very tight time lines. The following table (Table 2) shows the outage plan with just 26 hours allocated for a blind to blind per bed installation plan.

Event	Scheduled Time	Actual Time	Difference
Install Blinds	2.0 Hours	2.0 Hours	
N2 Blowdown	2.0 Hours	2.0 Hours	
Open Top Manway	1.0 Hours	0.75 Hours	0.25 Hours
Obtain Entry Permit / Safety review Meeting	1.0 Hours	1.0 Hours	
Equipment Rigging	2.0 Hours	2.0 Hours	
Unloading Bed	3.0 Hours	2.0 Hours	1.0 Hours
Vessel Cleaning	1.0 Hours	0.5 Hours	0.5 Hours
Remove Outlet Basket	2.0 Hours	0.5 Hours	1.5 Hours
Install OSG	4.0 Hours	1.5 Hours	2.5 Hours
Loading Molecular Sieve Bed	4.0 Hours	2.5 Hours	1.5 Hours
Close Manways / Un-install Blinds	4.0 Hours	3.5 Hours	0.5 Hours
Total Hours **Table 2.**	**26 Hours**	**18.25 Hours**	**7.75 Hours (Schedule Positive)**

The Outlet Support Grid is designed and installed without the need to weld to the vessel, no rope packing, no sealing materials and no thermal blankets are required to form a reliable sealing solution to the vessel head. The Outlet Support Grid mechanical sealing system preforms extremely well over the vessel and process thermal cycles which is critical in the full bed containment solution preventing media migration.

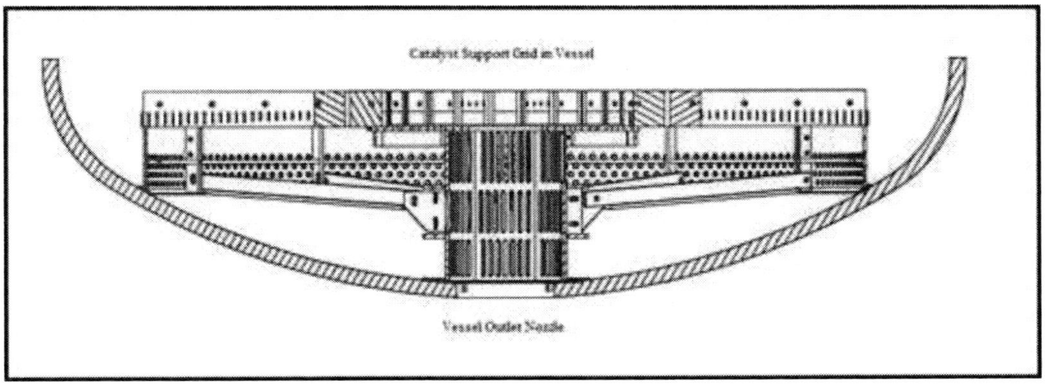

Installed View

Given the safety concerns of the high ΔP of 130 PSIG and rising, Woven Metal Products Outlet Support Grid was engineered to address these concerns and meet the bed design pressure drop of 3.08 PSIG at startup. With properly designed profile wire applied on all grid surfaces, the Outlet Support Grid has shown and will continue to perform at a very low and predictable ΔP as shown in table one since start-up. The Outlet Support Grid is designed to be in direct contact with the molecular sieve or any other media as the profile wire slot spacing is designed for the catalyst size loaded into the bed. Any support media utilized further assists in the bed hydraulics and should be designed and loaded for this purpose. As loaded the following inert balls were utilized 1/2" (bottom), ¼" (mid layer) and 1/8" (top layer) in contact with the molecular sieve, each application should be carefully examined and designed by catalyst providers.

Post – Installation Bed Loading			
Media	**Pounds**	**Cubic Feet³**	**Feet / Inches**
Ceramic Inert Balls ½"	828 (estimate)	19	0.5 Feet
Guard Bed Media	2,200	50	1.8 Feet
4 x 8 mesh 1/8" Molecular Sieve	10,500	228.26	6.1 Feet
8 x 12 mesh 1/16" Molecular Sieve	6,000	127.66	3.98 Feet
Ceramic Inert Balls 1/8" (Bottom – Top)	414 (estimate)	10	0.25 Feet
Ceramic Inert Balls ¼" (Bottom – Middle)	414 (estimate)	10	0.25 Feet
Ceramic Inert Balls ½" (Bottom – Bottom)	414 (estimate)	15	0.25 Feet
Totals	20,770 (estimate)	459.92	13.13 Feet

The Midstream company's operations utilize a timer method for bed cycling; previous timer cycles were set at adsorption for 12 hours and regeneration for 6 hours. Post installation adsorption cycle is now set at 16 hours and regeneration is set at 6 hours. With the molecular sieve bed hydraulics optimized, the bed timers are being increased steadily over time while watching for potential bed break through. The initial bed performance indications and the plant records for these dehydration beds show minimal filter dusting after one year of continuous processing. These are all positive indications that these molecular sieve beds will meet or exceed the estimated 3 years of bed life giving the Midstream companies operations a much safer and very predicable operation.

This project and the installation were considered very successful with the dehydration vessel and the bed support concerns addressed and corrected including;

- Maximum volume of molecular sieve for the unit
- Longer molecular sieve bed adsorption cycles which enhance throughput / plant de-bottlenecking
- Fewer bed regeneration cycles reduces the amount of molecular sieve breakdown over time
- Full bed containment
- Full bed utilization
- Consistent dew point control
- Predictable bed performance
- Reduce fuel gas consumption

Outside the vessel benefits can be examined and appreciated with full bed utilization, fewer bed regeneration cycles will be required. The operational impact of fewer bed regeneration cycles is easily calculated in the reduction of valve cycles – given each bed includes 6 inlet / outlet valves, incoming wet gas feed valves, steam and heater control valves essentially doubling the operational life of these valves greatly reducing the required maintenance costs for each valve. Each plant should consult with their valve manufacture / supplier to fully understand these operational valve benefits and the potential maintenance cost savings. Full bed containment aids not only in the valve life with less contamination present, full bed containment greatly reduces downstream filter plugging issues, cryogenic expander unit plugging and the need to bring the operational plant down, open and clean these units at un-scheduled times and loosing valuable production profits.

Conclusion

As with all plants, processes and capital projects many factors must be considered and evaluated. Return on investment (ROI) calculations can include many of the areas evaluated and discussed in this field installation and within this paper. Many factors will need to be evaluated including personnel and plant safety, natural gas prices, operational needs, operational costs, opportunity costs associated with lost production, lost efficiencies in molecular sieve beds and the installed equipment that are not engineered to maximize production capabilities and capacities. More information on the Outlet Support Grid and many other equipment solutions can be found at our website www.wovenmetal.com

513678 Regeneration Peak Profiles of Temperature Swing Adsorption Processes

Abstract:

Temperature Swing Adsorption (TSA) processes have been widely used for natural gas processing for removing water, sulfur, CO2, heavy hydrocarbon, etc. In a TSA process, a feed gas first enters an adsorber to adsorb the contaminants. The adsorber is regenerated by raising the temperature of the adsorbent typically by purging the bed with a preheated gas. The hot regeneration effluent gas, after cooling, is further processed to remove the contaminants before this spent regeneration gas can be reused, recycled or disposed. Due to the dynamic nature of the TSA process, the compositions in the effluent regeneration stream tend to vary with time, generating a peak concentration, several orders of magnitude higher than its average. The downstream unit to process the spent regeneration gas needs to be sized based on the peak, which results in a significant increase of the equipment size. The fluctuation of stream compositions can also result in unstable operation for the downstream units, especially if combustion or incineration is involved in these downstream processing units.

In this presentation, we will show that if the spent regeneration gas is to be recycled to the front end of a TSA unit, the TSA unit does not need to account for the peak and fluctuation of the inlet compositions. If the spent regeneration gas is to be processed in a solvent-based absorption unit, a fixed-bed adsorber can be used to dampen the peak concentration profile from the spent regeneration gas. The benefit of this solution will be demonstrated using an example from a Middle East natural gas complex, where UOPâ€™s Amine Guard^{TM} FS, Molecular Sieve, and Selexol^{TM} processes are used for pretreatment.

511840 CFD Simulation of a Plunging Jet Down-Flow Bubble Column with Variable Free-Jet Length Using a Euler-Euler Method

Abstract:

The key feature of a plunging jet down-flow bubble column is a plunging liquid jet that is used to entrain the inlet gas in a column of liquid. Experimental characterization of the plunging jet and the subsequent gas entrainment has been studied extensively and the physics is quite well understood [1,2]. But still the successful CFD simulation of:

1) the plunging liquid jet,

2) the associated gas entrainment in the mixing zone and

3) the resultant two-phase bubbly dispersion flow in the downstream section;

Capturing all these three phenomena in a single simulation still remains a challenge. Figure 1 shows the three distinct zones in the column where the above three phenomena occur respectively. The existing high resolution CFD models like the Volume Of Fluid (VOF) method and Front-tracking method have been successfully employed to capture the phenomena occurring above and below the free-liquid surface [3–5], but they are limited in application for cases where the entrained gas bubbles are of few mm size, and not suitable/practical for capturing the entrainment and dispersion of micro-bubbles. And further, as their computational requirement is very high, these models are not suitable for full-reactor scale simulations. On the other hand the low resolution model like Eluer-Euler model has been successfully employed to capture only the phenomena occurring below the free surface by applying a two-phase bubbly flow inlet with pre-determined jet-velocity profile from the free surface of the liquid pool [6–8]. This approach works fine only for cases were the tank or container cross-sectional area is sufficiently large and the free surface sloshing can be neglected.

In the presented work a sigmoid function based drag modification is used in a 3D Euler-Euler framework to capture both the complex gas entrainment process and also the two-phase gas-liquid bubbly flow in a single simulation. Equation 2 shows the formulation of the drag-modification function used, and Figure 2 shows the variation of the drag-modification factor as a function of the local gas holdup.

(1)

(2)

Figure 3 shows the mesh used for the CFD simulations. The successful implementation of the drag modification enabled the CFD simulations to reproduce the experimentally observed free jet lengths and the gas holdup in the two phase gas-liquid bubbly-flow section of the column. Figure 4 shows the comparison of the free-jet lengths obtained from CFD with the experimental data for the 14 different operating conditions considered. Figure 5 shows the difference in the transient evolution of the free-jet length with and without Drag-modification implementation, for one particular operating condition of $V_L = V_G = 29.39$ mm/s . Figure 5 clearly highlights the need for implementing the drag-modification, without which the CFD simulations are not able to capture the entrainment of the gas-phase into the liquid phase by the plunging liquid jet. Further, the high-resolution turbulence and holdup data obtained from the 3D CFD simulations for the different operating conditions were utilized to perform a linear stability analysis as derived by Ghatage et al. [9]. The results from the CFD simulations were key to calculating the stability parameter K_3 used in the linear stability analysis of Ghatage et al. The parameter K_3 is in turn a function of the turbulent kinetic energy (k) at the mixing-zone exit (Equation 3), so CFD is required for both locating the mixing-zone exit (form axial profiles of turbulent kinetic energy) and also for evaluating the value of K_3.

(3)

This value of K_3 serves as an input to the stability analysis, which in turn determines the critical gas holdup at which the regime transition from the homogeneous regime to the heterogeneous regime takes place. The prediction of this critical holdup at which the regime transition occurs is crucial for the design and scale-up of bubble column reactors.

Figure 1: Schematic of the plunging jet bubble column

Figure 2: Variation of the Drag modification function as a function of the local gas hold-up.

Figure 3: Mesh used for the CFD simulations

Figure 4: Comparison of the Free jet lengths obtained from the CFD simulations with the Experimentally observed values.

Figure 5: Comparison of the time evolution of the contours of gas hold-up, (a) without any Drag modification and (b) with drag modification implementation.

Figure 6: Axial variation of the parameter K_3 used in the linear Stability analysis (of Gatage et al.), evaluated using equation 3.

References

[1] A.K. Roy, B. Maiti, P.K. Das, Visualisation of air entrainment by a plunging jet, in: Procedia Eng., Elsevier, 2013: pp. 468–473. doi:10.1016/j.proeng.2013.03.148.

[2] K.T. Kiger, J.H. Duncan, Air-Entrainment Mechanisms in Plunging Jets and Breaking Waves, Annu. Rev. Fluid Mech. 44 (2012) 563–96. doi:10.1146/annurev-fluid-122109-160724.

[3] S.S. Deshpande, M.F. Trujillo, X. Wu, G. Chahine, Computational and experimental characterization of a liquid jet plunging into a quiescent pool at shallow inclination, Int. J. Heat Fluid Flow. 34 (2012) 1–14. doi:10.1016/j.ijheatfluidflow.2012.01.011.

[4] X.L. Qu, L. Khezzar, D. Danciu, M. Labois, D. Lakehal, Characterization of plunging liquid jets: A combined experimental and numerical investigation, (2011). doi:10.1016/j.ijmultiphaseflow.2011.02.006.

[5] L. Khezzar, N. Kharoua, K.T. Kiger, Large eddy simulation of rough and smooth liquid plunging jet processes, Prog. Nucl. Energy. (2015). doi:10.1016/j.pnucene.2015.06.011.

[6] E. Krepper, F.-P. Weiss, S. Alt, A. Kratzsch, S. Renger, W. Kästner, Influence of air entrainment on the liquid flow field caused by a plunging jet and consequences for fibre deposition, Nucl. Eng. Des. 241 (2010) 1047–1054. doi:10.1016/j.nucengdes.2010.04.002.

[7] F.Z. Kendil, D. V Danciu, M. Schmidtke, A.B. Salah, D. Lucas, E. Krepper, A. Mataoui, Flow field assessment under a plunging liquid jet, (2012). doi:10.1016/j.pnucene.2011.12.009.

[8] F.Z. Kendil, E. Krepper, A.B. Salah, D. Lucas, A. Mataoui, Numerical study of a bubble plume generated by bubble entrainment from an impinging jet, Nucl. Eng. Des. 241 (2011) 4111–4121. doi:10.1016/j.nucengdes.2011.06.053.

[9] S. V Ghatage, M.R. Bhole, N. Padhiyar, J.B. Joshi, G.M. Evans, Prediction of regime transition in three-phase sparged reactors using linear stability analysis, Chem. Eng. J. 235 (2014) 307–330. doi:10.1016/j.cej.2013.08.095.

513513 Design and Analysis of Cryogenic Air Separation Processes Integrated with Liquefied Natural Gas Gasification

Abstract:

Liquefied natural gas (LNG) is a major source of natural gas supply for long-distance
transport. After the long travel, the LNG is vaporized in a receiving terminal in order to supply natural gas at desired conditions through domestic pipeline networks. Due to the extremely low temperature LNG regasification plant can provide, ASU unit can benefit significantly from the cold exergy.

However, in the heat integration, highly flammable hydrocarbons in the presence of oxygen pose a risk. Advanced equipment fabrication techniques minimize the risk, however there is always a possibility of leak in the equipment and it is desirable to have inherently safer equipment.

The main object of the project is to analyze the conventional integrated Heylandt system using inherent safety index. Based on the safety analysis, stage to identify the equipment which needs improved design. Two designs are proposed which minimize the hazards by introducing an intermediate fluid for the heat transfer. Mixed Refrigerant (MR) is used as the heat transfer fluid in Design 1. Propane is used as the heat transfer fluid in Design 2. A comparative safety analysis of the proposed designs is done, indicating that the new designs are safer as compared to the conventional design.

511160 Numerical Study of Bund Overtopping after Catastrophic Tank Failures

Abstract:

Gas and petroleum products are stored in a large quantity using tanks. Impoundment systems, such as bund or dike, are put in place to control the spill in case of a failure of primary containments. A catastrophic tank failure, although not likely, can cause an unacceptable consequence as demonstrated in the previous incidents. Nowadays, natural gas is widely used as a feedstock in chemical industry and a fuel to generate power, since it is abundant in terms of reserve and a cleaner energy compared with other fossil fuels. NFPA 59 A specifies the requirements of constructing Liquefied Natural Gas (LNG) storage tanks and dikes, which requires the dike to contain 110% of tank capacity. Previous works using water, liquid nitrogen and LNG have approved that this criteria is insufficient for a dike to contain all the spillage in case of a catastrophic tank failure. Simple correlations were proposed to predict overtopping ratios by regression against experimental data, however, the physical mechanism remains unclear for the behaviors of fluid flow and the interaction of flowing material with dike wall. This work aims to develop a Computational Fluid Dynamics (CFD) model to understand the mechanism of bund overtopping. The model, which will be validated against experimental results, can contribute to conducting a risk assessment of bund overtopping events and provide recommendations for dike design and installation.

Keywords: *Tank failtures, Bund ovetopping, CFD, Consequence analysis, LNG*

Liquefied Natural Gas (LNG) Pool Spreading and Vaporization on Water

Nirupama Gopalaswami[1],
Konstantinos Kakosimos[2],
Bin Zhang[1],
Yi Liu[1],
R. Mentzer[1]
and
M. Sam Mannan[1*]

[1]Mary Kay O'Connor Process Safety Center
Artie McFerrin Department of Chemical Engineering
Texas A&M University System

College Station, Texas 77843-3122, USA
[2]Mary Kay O'Connor Process Safety Center - Qatar
Texas A&M University at Qatar, PO Box 23874, Education City, Doha, Qatar
*Corresponding author: +1-(979) 862-3985; mannan@tamu.edu

Prepared for Presentation at
American Institute of Chemical Engineers
2018 Spring Meeting and 14th Global Congress on Process Safety
Orlando, Florida
April 22 – 25, 2018

Liquefied Natural Gas (LNG) Pool Spreading and Vaporization on Water

Abstract

The investigation of pool spreading and vaporization phenomenon is an essential part of consequence analysis to determine the severity of LNG spills on water. In this study, release of LNG on water during marine operations is studied through experimental and numerical methods. The study involves emulation of an LNG leak from transfer arms during side by side loading operations. The experimental part involves flow of LNG in a narrow trench filled with water and subsequent measurement of pool spreading and vaporization parameters. The numerical part involves CFD simulation using a three dimensional hybrid homogenous Eulerian multiphase solver to model the pool spreading and vaporization phenomenon. In this method, LNG is modeled as dispersed phase droplets which can interact with continuous phases – water and air through interphase models. The numerical study also employs a novel user-defined routine for capturing the LNG vaporization process. The CFD solver was capable of capturing the salient features of LNG pool spreading and vaporization phenomena. It was observed from experiment and CFD simulation that wind influenced both pool spreading and vaporization phenomenon through entrainment and convection.

Keywords

CFD;
Liquefied natural gas;
LNG;
Pool spreading;
Vaporization;

1. Introduction

The development of large natural gas fields in remote locations is currently being exploited through advancement in technology. A marked increase in LNG utilization is forecast with an increase in LNG marine operations like transportation, bunkering and loading. A release of LNG on water during these marine operations can occur due to several factors like collisions, grounding, loss of mechanical integrity of flanged connections in loading arms, inadvertent disconnection of hoses, brittle fracture of materials lining the loading arms, overfilling or over-pressurizing of fuel tanks, undesirable ship maneuvering due to human factor issues and extreme weather scenarios like hurricanes [1]. Upon release, a spreading liquid can form a pool with rapid vaporization, leading to the formation of a flammable vapor cloud. One of the key steps in safety analyses involves the determination of maximum extent of the pool and the rate at which flammable vapor is being produced.

Given the relevance of this research, a number of researchers have already studied LNG releases on water by both experimental and numerical means [2],[3]. The pool spreading parameters like pool area, height and spreading rate were measured dynamically by recording the distance covered by the pool in small scale apparatus [4]. In small scale experiments, LNG was impeded by ice formation, which led to under-prediction of pool spreading parameters. Contrarily, it was also observed that these parameters were difficult to measure in medium and large scale experiments due to vapor blocking [5]. Similar to pool spreading parameters, the vaporization rate in LNG experiments was determined through three different methods. The first method involved measurement of mass loss due to vaporization and subsequent determination of vaporization rate using the slope of mass loss data [6]. The second method involved measurement of water temperature and application of empirical correlations on heat transfer mechanisms like convective boiling or conduction to obtain the vaporization rate [7]. The third method involved measurement of vapor flow through an array of gas sensors and integration of measured concentration data across the area covered by gas sensors [8].

Over the years, efforts were also directed to develop numerical models to simulate these physics. The approaches can be classified as phenomenological, integral, shallow layer and Computational Fluid Dynamics (CFD) models. The phenomenological models provide hypothesized relationship for pool spreading and vaporization based on the forces and heat sources acting on the pool. (*e.g.* [9],[10]). Complex physics like hydraulic jump and flashing of LNG droplets could not be modeled using phenomenological models. The integral model calculates the dynamic behavior of pool spreading and vaporization through mass and energy balances (e.g. PHAST [11]). The integral models vary based on the assumptions made for spreading rate (uniform pool thickness), heat transfer models (conduction, convection), release types (continuous vs instantaneous) and the numerical scheme used for solving the equations (second order, first order discretization). The integral models have been validated against many LNG experiments [12]. Scenarios involving obstacles like process equipment or transport systems cannot be modeled using integral models. The Shallow layer model assumes that the transport properties of the fluid vary along the lateral directions, and is constant along elevation (*e.g.*[13]). The shallow layer model fails when there is significant front resistance due to partial submergence of LNG on water, as water is modeled as a wall in these models rather than a fluid. Inclusion of turbulence due to bubbly vaporization also poses significant difficulty as three dimensional solvers are required for turbulence models. A recent improvement in consequence modeling is the use of CFD solvers. Although, CFD simulations are computationally expensive and analytically challenging, they can replicate complex multiphase physics in obstacle laden environments.

It is the aim of this research to address these gaps in experimental and numerical methods for determining the pool spreading and vaporization parameters of LNG released on water. Particular attention has been paid to specific release scenario like a LNG leak between two ships during side by side loading operation for which data is currently lacking. An experimental study was conducted by allowing LNG to flow through a narrow channel of water. Additionally, a numerical study was performed to simulate the pool spreading and vaporization behavior. Finally, the CFD results are validated against the experimental results.

2. Materials and Methods

2.1. Experimental Methodology

The experiment was performed in an L-shaped trench present in Brayton Fire Training Field (BFTF), College Station. Each leg (leg 1 and leg 2) of the trench is 8.2 m long, 1.22 m wide and 1.05 m deep (see Fig.1). Prior to the experiment, the trench was filled with water up to a height of 0.8m. A weather station was installed near the trench to obtain the local conditions such as wind speed, wind direction, atmospheric pressure, temperature, and relative humidity. The trench was equipped with N-type thermocouples in a linear arrangement to measure the water temperature and LNG temperature during the spill. The tip of the thermocouples were above water for measuring LNG temperature and below water for measuring water temperature. Table 1 provides the locations of thermocouples with respect to origin. The bottom left corner of leg 1 was considered as origin. The experiment was recorded with three video cameras and one Infra-Red (IR) camera. Cameras 1 and 2 were placed to track the pool spreading of LNG in leg 1 and leg 2 respectively and camera 3 provided an aerial view of the experiment. An ultrasonic level transmitter (Echopod DL-24) was placed in leg 2 to track the dynamic height of LNG during the vaporization process. The thermocouples' and level sensors' output were recorded every second (1Hz) by a Data Acquisition system (DAQ). A summary of the release conditions and atmospheric conditions present during the experiment is provided in Table 2.

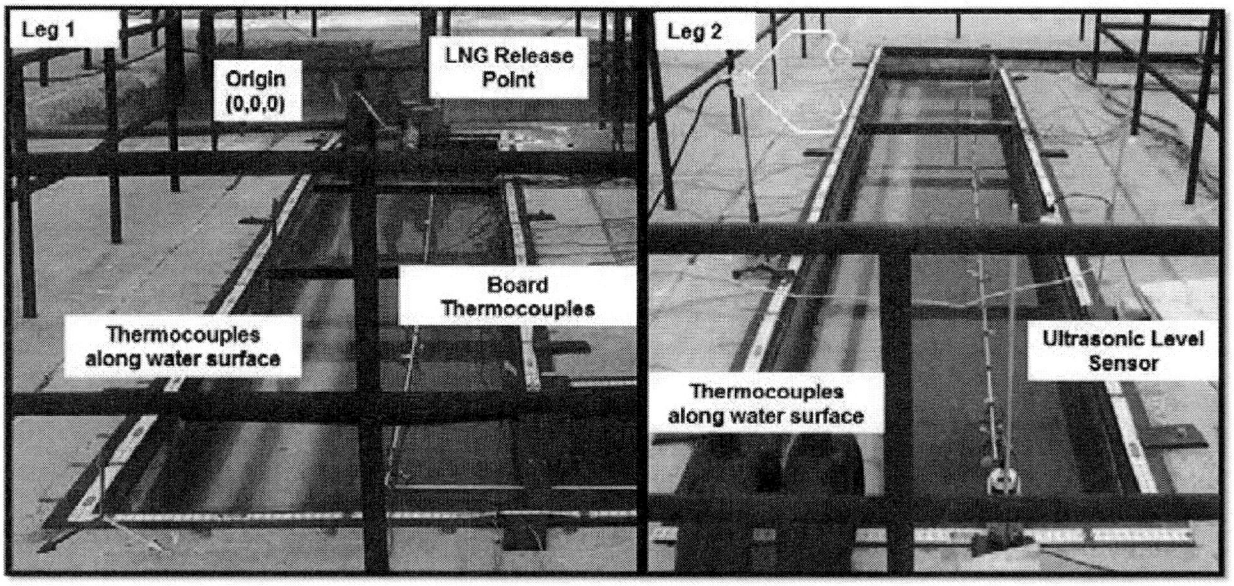

Fig. 1. Experimental setup

LNG was continuously discharged from a tanker of 11000 gallon capacity through a cryogenic hose (internal diameter-0.076m). The discharge point is present at the surface of Leg 1 above origin (see Fig 1). The LNG height was monitored in the tanker and this value was used to determine the volume lost in tanker with respect to time and subsequently the flow rate of LNG. As LNG was released into the trench, the contact area between LNG and water increases steadily as it spreads on water. This area covered by LNG was determined by recording the time at which the thermocouples record LNG boiling temperature. As the locations of the thermocouples were known prior to the experiment, the area was then determined by multiplying the length and the width from the origin. It is assumed that LNG covers the entire width of the trench when it touches the thermocouples. The pool area during regression was obtained experimentally by recording the time at which water temperature returns to atmospheric temperature during the experiment. The pool spreading and regression of LNG on water during experiment is shown in Fig.2. The spreading rate is determined in a similar manner by calculating the time taken by LNG to reach selected thermocouples present in the experiment. The spreading rate is then determined as distance by time. This method gives an estimate of spreading rate when LNG touches the thermocouples for the first time. It was also observed from previous experiments that pool height strongly affected the vaporization rate of LNG [14]. Hence pool height was measured using three different methods. In the first method, the pool height was measured manually using a dip stick at regular intervals in the intersection of leg 1 and leg 2 (Fig.3 (a)). The second method involved a linear array of thermocouples in the intersection of leg 1 and leg 2 to determine the pool height (see Fig.3 (b)). As LNG level started to increase, the thermocouples array measured the LNG temperature. Since the positions of thermocouples are known prior to experiment, pool height is calculated by recording the thermocouples which registered LNG temperature. As the pool reached the end of leg 2, the pool started to regress. The pool height was captured dynamically using ultrasonic level sensor (Fig.3(c)). In this experiment, both mass loss measurement using weighing balance and gas detectors were not employed and hence, the vaporization mass flux was determined using temperature values. The temperature of water and LNG was measured and empirical correlations (equations (19)-(25)) on convective heat transfer and boiling was applied with water temperature to obtain the vaporization mass flux. This experiment will be referred to as "MKOPSC Trench experiment" in the rest of the manuscript.

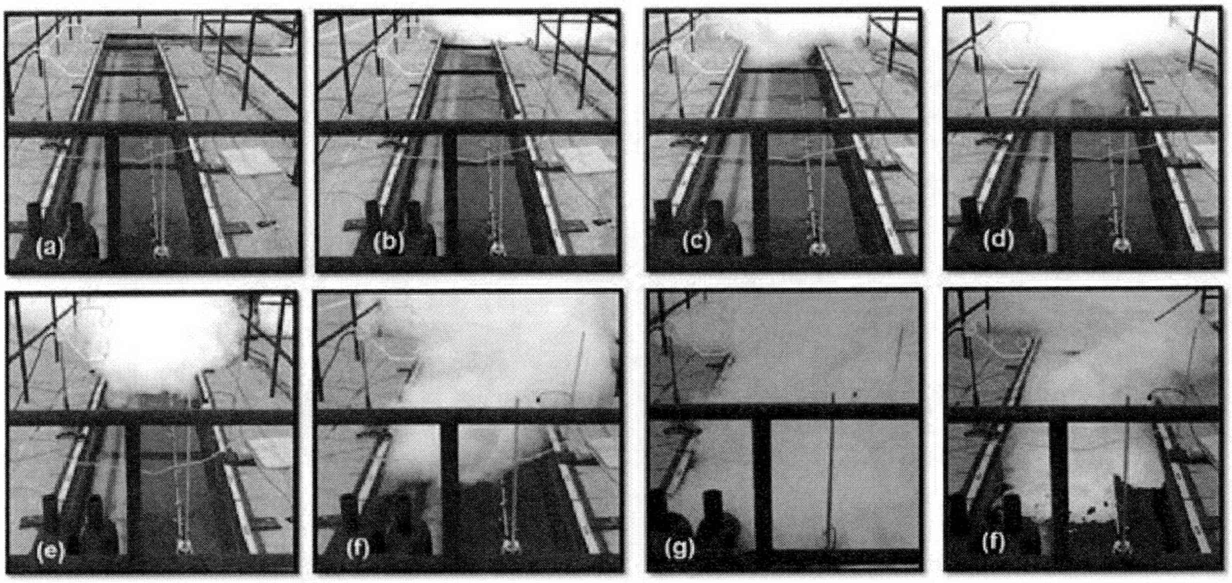

Fig. 2. Snapshots of LNG pool (a) traversing leg 1 of trench (b) entering the leg 2 of trench (c) covering 1/8th of leg 2 (d) covering 1/4th of leg 2 (e) covering ½ of leg 2 (f) covering ¾ th of leg 2 (g) entire trench (f) regressing due to vaporization

Table 1. Thermocouple Locations

Origin (0,0,0)					
Leg 1 Width from Origin-0.89 m			Leg 2 Width from Origin-0.91m		
TC No.	Length of TC from Origin (m)	Height of TC from Origin (m)	TC No.	Length of TC from Origin (m)	Height of TC from Origin (m)
TC-1	0.89	0.9	TC-14	9.73	0.88
TC-2	1.19	0.76	TC-15	10.03	0.75
TC-3	1.50	0.9	TC-16	10.34	0.84
TC-4	2.09	0.76	TC-17	10.64	0.76
TC-5	2.72	0.9	TC-18	10.97	0.83
TC-6	3.35	0.76	TC-19	11.26	0.76
TC-7	3.61	0.91	TC-20	11.55	0.86
TC-8	4.04	0.79	TC-21	11.86	0.76
TC-9	4.34	0.89	TC-22	12.47	0.85
TC-10	4.66	0.76	TC-23	12.77	0.76
TC-11	5.23	0.93	TC-24	13.08	0.85

TC-12	5.84	0.78	TC-25	13.38	0.76
TC-13	7.06	0.98	TC-26	13.69	0.85
			TC-27	13.98	0.76
			TC-28	14.3	0.85
			TC-29	14.6	0.78
			TC-30	14.91	0.85
			TC-31	15.21	0.76
			TC-32	15.52	0.88

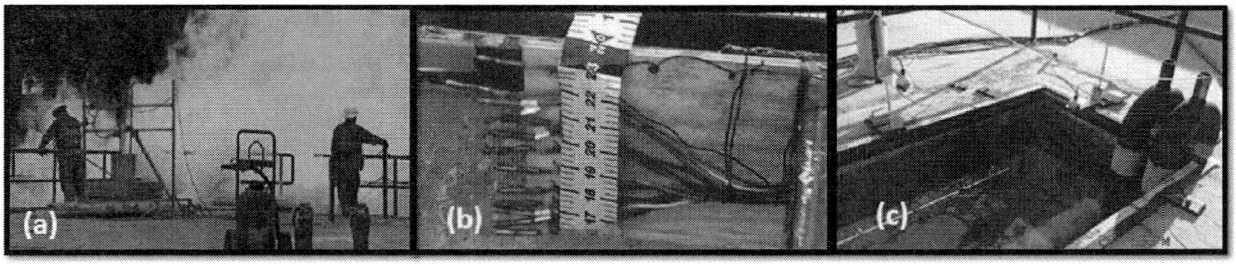

Fig.3. Pool height measurement through (a) dipstick (b) thermocouples (c) ultrasonic level sensor

Table 2. Summary of release conditions and atmospheric conditions

Release Conditions		
LNG flow rate	9.5±0.08 kg/s	
Spill duration	1200 s	
Water temperature	20 °C	
Spill area	17.8 m^2	
LNG Composition	100% Methane	
Height of discharge	0.1 m	
Diameter of discharge pipe	0.076 m	
Atmospheric Conditions		
Wind velocity	6.2±1.3	m/s
Wind direction	SE	Global
Atmospheric temperature	25.6±0.03	°C
Relative humidity,	52.6	%
Dew point temperature	15.1	°C
Stability class	C	
Solar radiation	4.34	$kW/m^2/day$
Sensible heat flux	18.06	kW/m^2
Ground Roughness length	3	mm [28]
Water surface roughness	1	mm [28]

2.2. Uncertainty Analysis

The pool spreading and vaporization parameters were primarily obtained from thermocouples. The uncertainty associated with the thermocouples can be due to several factors like the type of thermocouple, response time and data retrieval interval. The uncertainty can be quantified by quantifying the error in these parameters. The tolerance value of N-type thermocouples is ±1.7 K (iconel alloy - class 3 type- working range −270 to 1300°C). The response time for this type of thermocouple is 0.25s(Δt_r). The thermocouples are connected to Data Acquisition System which can record data at the speed of 1s (Δt_d).

Additionally, the uncertainty propagates when the temperatures values are used to determine pool spreading and vaporization parameters. The uncertainty in any experimentally derived variable (f_e) (pool spreading/vaporization parameter) is then a function of tolerance value, response time, data retrieval time and temperature. The error propagation is determined using Kline McClintock Method [15] expressed as

$$\frac{\Delta f_e}{f_e} = \sqrt{\left(\frac{\partial f_e}{\partial n}\right)(\Delta n)^2 + \left(\frac{\partial f_e}{\partial t}\right)(\Delta t_r)^2 + \left(\frac{\partial f_e}{\partial t}\right)(\Delta t_d)^2 + \left(\frac{\partial f_e}{\partial \Delta T}\right)(\Delta T)^2} \quad (1)$$

The propagated error is expressed as error bars in experimentally determined parameters (Fig 8, 10, 11 and 12).

2.3. Numerical Methodology

The computational domain consists of the L-shaped trench and a rectangular region to simulate the atmosphere above the trench (see Fig.4).

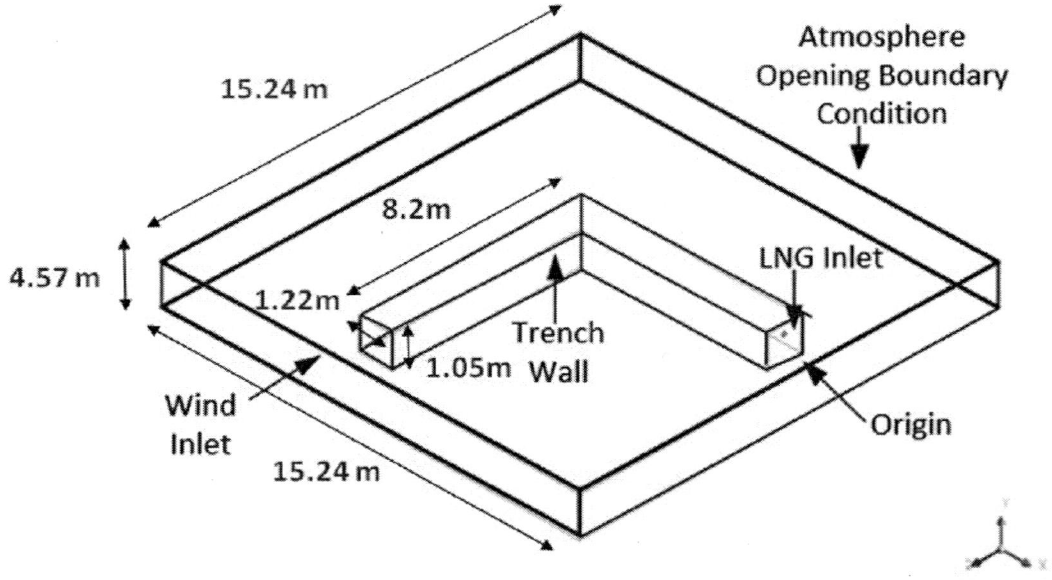

Fig. 4. Computational domain

2.3.1. Grid Sensitivity Analysis

The computational domain was divided into tetrahedral cells. A grid sensitivity analysis was performed by considering three different grids (see Table 3). All the other simulation parameters were kept identical to ensure that the results were changing as a result of the grid characteristics alone. Fig.5 shows the water temperature with respect to the distance from the LNG discharge. The water temperature is influenced by the discharge location and reduced more near the discharge when compared to regions remote from the discharge location. As can be seen in Fig.5, the results with grids 2 and 3 are almost identical, whereas grid 1 is different in magnitude and trend. A grid-independent solution was achieved with grid 2 consisting of 3.2×10^6 elements. The element size in this grid was restricted to 0.05m in the trench region and 0.1m in the atmospheric region to allow for the optimum resolution of interface between water-LNG and LNG-air phases.

Table 3. Grid sensitivity analysis

Parameters	Grid 1	Grid 2	Grid 3
Smallest element size	0.075 m	0.05 m	0.05 m
Largest element size	0.15 m	0.1 m	0.075 m
Size difference between smallest and largest element	0.075 m	0.05 m	0.025 m
Total Nodes	176154	573256	1346905
Total Elements	969218	3245728	7706835
Total CPU Time	68080 s	262200 s	691200 s
Water temperature at 600 s	290.8 K	291.4 K	291.4 K

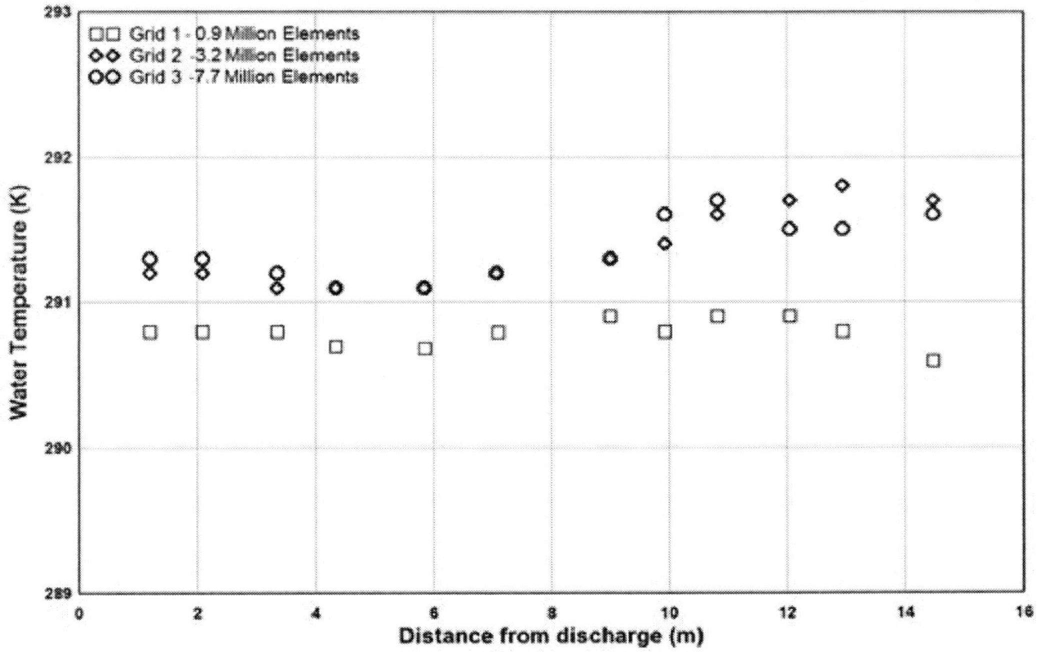

Fig. 5. Grid sensitivity analysis showing water temperature with respect to distance from discharge at t-600s.

2.3.2. Governing Equations

To model the LNG flow on water the homogeneous Eulerian multiphase model was adopted. This is a limiting case of Eulerian-Eulerian multiphase flow where all fluids share the same velocity, pressure, temperature and turbulence fields. The transport properties are solved in terms of phasic quantities. Alternate approach involves the multifluid Volume of Fluid (VOF) model [16], but it faces scale-up issues. The release of LNG on water involves a fragmented jet where small droplets vaporize and large droplets aggregate to form a pool [17]. To address this phenomenon, three components were defined- LNG, water and air. LNG was modeled as dispersed phase droplet with a mean droplet size of 0.017m. Air and water was defined as continuous phases. Currently there is lack of information regarding LNG droplet sizes. The droplet formation is part of discharge modeling which requires detailed study and is out of scope of this paper. The mean diameter was then based on the size range of bubbles (0.01-0.018m) observed in pure methane [18] and light LNG [19] during the vaporization process. Sensitivity analysis was performed on the droplet sizes 0.01m and 0.02m. Any value within this range provides similar results. Values lower than 0.01m increase the computational time, whereas values greater than 0.02 m are too large to be discretized with the current grid configurations. LNG was modeled as 100% methane.

2.3.2.1. Interphase Model

The three interfaces LNG-air, LNG-water and air-water are modeled using interphase models. The Particle model is selected for interfacial transfer between LNG-air and LNG-water. These two interfaces are constantly evolving with respect to time. The free surface model is selected to model the water-air interface. The water–air interphase is distinct and effects of water movement like ripples and waves are captured well using the free surface model. The interfacial area density (surface area per unit volume) is determined based on the assumption that LNG is present as spherical particle of mean diameter. The interfacial area density 'A' of each interphase is given by

Particle Model:

$$A_{LNG|Air} = 6\frac{r_{LNG}}{D_{LNG}} \tag{2}$$

Particle Model:

$$A_{Water|LNG} = 6\frac{r_{LNG}}{D_{LNG}} \tag{3}$$

Free Surface Model:

$$A_{Water|Air} = \frac{2\,|\nabla r_{water}||\nabla r_{air}|}{|\nabla r_{water}|+|\nabla r_{air}|} \tag{4}$$

2.3.2.2. Mass Transport

The mass transfer of LNG between different phases is governed by the continuity equation [20] expressed as

$$\frac{\partial \rho}{\partial t} + \nabla.(\rho U) = mA \quad (5)$$

The transport properties in terms of phasic fraction is expressed as

$$\rho = r_{LNG}\rho_{LNG} + r_{air}\rho_{air} + r_{water}\rho_{water} \quad (6)$$

$$mA = m_{LNG|air}A_{LNG|air} + m_{LNG|water}A_{LNG|water} + m_{air|water}A_{air|water} \quad (7)$$

$m_{LNG|Water} = m_{Air|Water} = 0$ as there is no mass transfer from LNG to water or air to water phase. However, there is transfer from LNG to air phase through vaporization. Eqn (5) now becomes,

$$\frac{\partial \rho}{\partial t} + \nabla.(\rho U) = m_{LNG|air}A_{LNG|air} \quad (8)$$

The mass transfer from LNG to air phase $m_{LNG|air}$ is modeled using CFX expression language using eqn (25). Here, the mass conservation equation for the mixture is obtained by summing the mass of three components alone (LNG, water and air).The non-zero source term in RHS of eqn (8) denotes the production of LNG vapor. The production of LNG vapor is assumed to be rapid and it escapes the computational domain immediately upon production. Hence LNG vapor is not included in the mixture mass fraction nor as a separate phase.

The volume fraction is solved using the volume conservation equation. The volume conservation equation is the obtained by dividing the mass conservation equation by phasic density and summing for all phases on both side of the equation,

$$\sum_{r=1}^{3}\left(\frac{1}{\rho}\right)\left(\frac{\partial \rho}{\partial t} + \nabla.(\rho U)\right) = \sum_{r=1}^{3}\left(\frac{1}{\rho}\right) m_{LNG|air}A_{LNG|air} \quad (9)$$

The volume fraction of water and air were specified as initial conditions. The volume fraction of LNG and change in volume fraction of all water and air are solved using eqn (9).

2.3.2.3. Momentum transport

The momentum transfer between the phases is expressed as -

$$\frac{\partial(\rho U)}{\partial t} + \nabla(\rho UU) = -\nabla p + \nabla.\tau + \rho g \quad (10)$$

$$\tau = r_{LNG}\tau_{LNG} + r_{air}\tau_{air} + r_{water}\tau_{water} \quad (11)$$

$$p = r_{LNG}p_{LNG} + r_{air}p_{air} + r_{water}p_{water} \quad (12)$$

For Homogeneous model, velocity fields are shared between the components

$$U_{LNG} = U_{air} = U_{water} = U \qquad (13)$$

The interfacial terms cancel out resulting in a single transport equation with variable density.

2.3.2.4. Turbulence Model

The initial growth of vapor cloud is characterized by a large magnitude of turbulence due to vaporization. This turbulence was caused due to convection currents called 'thermals' which rise from the water-LNG interface and break in the LNG-air interface [14]. The turbulence was modeled using the standard k-ε turbulence model [21] in Ansys CFX with default values for constants. The values of turbulence kinetic energy and eddy dissipation rate for cryogenic liquid boiling on water were obtained from a high speed flow visualization experiment [14]. The results provided values of turbulent kinetic energy ($0.005 m^2/s^2$) and eddy dissipation rate ($0.004\ m^2/s^3$) which were provided as inputs to the turbulence models.

2.3.2.5. Energy Transport

The energy transport is given by

$$\frac{\partial}{\partial t}(\rho E) + \nabla.(\rho U E) = \nabla.(k\nabla T) + \tau{:}\nabla U + \dot{Q} \qquad (14)$$

$$E = r_{LNG}E_{LNG} + r_{air}E_{air} + r_{water}E_{water} \qquad (15)$$

$$k\nabla T = k_{LNG}\nabla T_{LNG} + k_{air}\nabla T_{air} + k_{water}\nabla T_{water} \qquad (16)$$

$$\dot{Q} = A_{LNG|air}q_{LNG|air} + A_{water|air}q_{water|air} + A_{LNG|water}q_{LNG|water} \qquad (17)$$

The heat transfer between water to air is negligible when compared to heat transferred between LNG and water and LNG and air phases. Hence,

$$q_{water|air} \sim 0 \qquad (18)$$

When heat is being transferred from water to LNG, the water temperature declines with time whereas the LNG temperature stays at its boiling point. From the previous experiments, it was observed that LNG stays in film boiling when released on water [22]. Due to this reason, the Berenson model for film boiling was selected for determining the heat transfer coefficient [23]. According to Berenson's model [23], the heat flux is a combination of temperature difference and heat transfer coefficient expressed as

$$q_{LNG|\,water} = h.\Delta T = 0.425(T_{water} - T_{LNG}).\left[\frac{k_V^3 \Delta H_V \rho_V(\rho_{LNG}-\rho_V)}{\mu_V \Delta T\left(\frac{g(\rho_{LNG}-\rho_V)}{\sigma_{LNG}}\right)^{\frac{1}{2}}}\right]^{\left(\frac{1}{4}\right)} \qquad (19)$$

Similarly the heat transfer from air to LNG is convective in nature, provided by

$$q_{LNG|air} = h_{air}.(T_{air} - T_{LNG}) \tag{20}$$

The heat transfer coefficient, h_{air} is given by

$$h_{air} = \frac{Nuk_{air}}{L} \tag{21}$$

Where the Nusselt number, Nu is given by

$$Nu = 0.037\ Re^{0.8}Pr^{0.33} \tag{22}$$

Reynolds number, Re is given by

$$Re = \frac{Lv_{air}\rho_{air}}{\mu_{air}} \tag{23}$$

And Prandtl number, Pr is given by

$$Pr = \frac{Cp_{air}\mu_{air}}{k_{air}} \tag{24}$$

The mass vaporization flux of LNG to air phase ($m_{LNG|air}$) is then obtained from energy balance as

$$m_{LNG|air} = \frac{\dot{Q}}{\Delta H_V} \tag{25}$$

Equation (19)-(25) were previously validated with experimental data involving continuous releases of cryogenic liquid [22].

2.3.3. Simulation Parameters

The walls of the trench were provided with an adiabatic no slip boundary condition. The LNG pipe of diameter 0.076m was provided as inlet by specifying the bulk mass flow rate. The head space above the trench was open to atmosphere where LNG and air could flow into and out of the domain. A wind velocity was provided at the wind inlet to simulate the wind effects on the pool (see Fig.4(a)). The pool was found to be influenced only by the wind above the pool. Hence a uniform wind velocity was used in this study. The initial height and volume fraction of water and air was defined using the step function in CFX.

The simulation was solved in Linux HPC cluster with 16 parallel processors and 16 CPU cores. The criterion of convergence of Navier Stokes equation was set to $1E^{-4}$. The time-stepping was adaptive providing an automatic control of time-step to achieve convergence in transient run. A list of salient inputs of the simulation is given in Table 4. The total computational time was 80 hours. This CFD methodology will be referred to as 'LSPREAD' simulation (LNG pool spreading and vaporization), in the rest of the manuscript.

Table 4. Salient inputs of LSPREAD methodology

Boundary Conditions	**Specification**
Trench wall	No-slip
LNG mass flow inlet	9.5 kg/s at 111 K
Wind Inlet	6.5 m/s at 293 K
Atmosphere	Opening at T=293 K and P=1 atm
Operating Conditions	**Specification**
Reference values	Reference pressure - 1 atm Reference density - 1.225 kg/m^3 Acceleration due to gravity- 9.8 m/s^2
Fluids	LNG, water and air
Models	**Specification**
Turbulence Model	k-ε [21]
Interphase Model	LNG-Air and LNG-Water- Particle model Air-Water - Free surface model
Heat Transfer Model	User-Defined Routine
Mass Transfer Model	User-Defined Routine
Solution Methods	**Specification**
Time-Stepping	Adaptive Minimum timestep- 0.001s Maximum timestep - 1s Coefficient loops - 10
Advection	Second-order scheme
Time derivative	Second-order backwards Euler
Volume Fraction	Coupled solver

3. Results

The results from LSPREAD simulation were validated against the experimental results. Experimental observations are provided in sections 3.1 and 3.2. Validation of key parameters like pool area, spreading rate, pool height and vaporization mass flux are demonstrated in sections 3.3-3.6.

3.1. Flashing of LNG

When LNG is released from a pipe, it flashes initially as LNG gets heated by the walls of the pipe before it contacts the atmosphere. Fig.6 compares the vapor emanating from the LNG pipe during the initial stages of the spill using (a) normal camera, (b) IR camera and (c) CFD simulation. LNG was found to be flashing for duration of 30s producing vapor and LNG droplets. The concentration of vapor during flashing was measured previously in field experiments and was typically low of the order of 2-3% v/v [24]. The visible boundary of vapor was also observed from IR camera and the temperature of vapor was approaching the atmospheric temperature during flashing. While the LNG vapor cloud has a distinct white appearance due to condensation of water droplets in air, the LNG vapor cloud during flashing was relatively translucent providing a distinct variation in the visible appearance between flashing and vapor cloud formation. In MKOPSC trench experiment and LSPREAD simulation, the height of the methane vapor extended up to the fence surrounding the trench (~1m). With further continuation of LNG spill, a dense vapor cloud was formed with significant increase in vapor cloud volume. The restriction of the spill to the boundaries of the trench was well captured by the LSPREAD simulation.

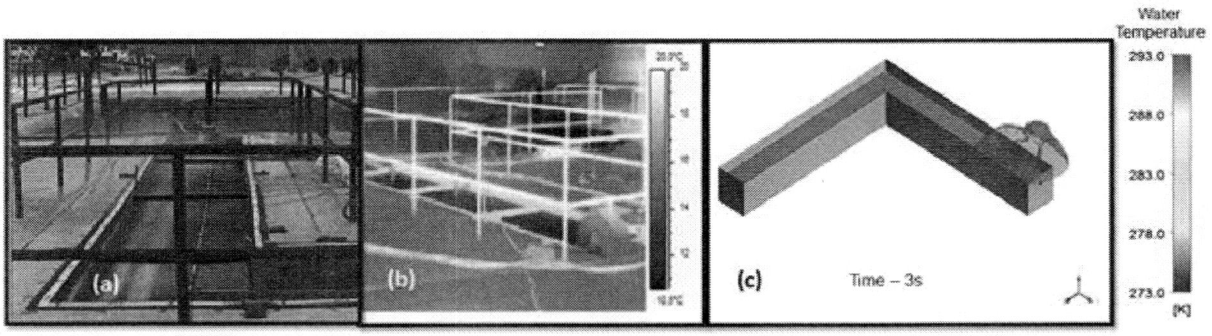

Fig. 6. Comparison of flashing phenomenon from (a) normal camera (b) IR camera (c) CFD simulation at t-3 s

3.2. Temperature Profiles

Fig.7. shows water temperature profiles of leg 1 and leg 2 thermocouples respectively. The thermocouples that were placed closer to the water surface underwent a large temperature gradient. An overall reduction in bulk water temperature of 4-14°C was observed in the experiment. This is an accordance with a similar study where liquid nitrogen was found to reduce the temperature of water to about 5-10°C during a continuous spill [25]. Similar reduction of temperature up to 8°C was obtained in the Falcon LNG tests [26]. It was also observed that the water temperature in leg 1 reduced more than leg 2. This is due higher contact time of LNG with water in leg 1 when compared to leg 2. Moreover, higher spreading rate in leg 1 contributed to steep temperature reduction in leg 1. Higher spreading rate in leg 1 lead to increased contact area and heat transfer. The temperature profiles from experiment were also compared to temperature profiles from CFD (Fig.7). As the LNG spread on water, significant amount of heat transfer occurred at the water surface. The water temperature from CFD was found to match well up to duration of 600 s in leg 1. After 600 s, an under-prediction was obtained by CFD. This is due to formation of ice layer, which hinders the temperature measurement of water below ice. The water temperature of leg 2 from CFD was similar in trend compared to experimental value with a small difference. This implied that the heat transfer was higher in CFD when compared to experiment.

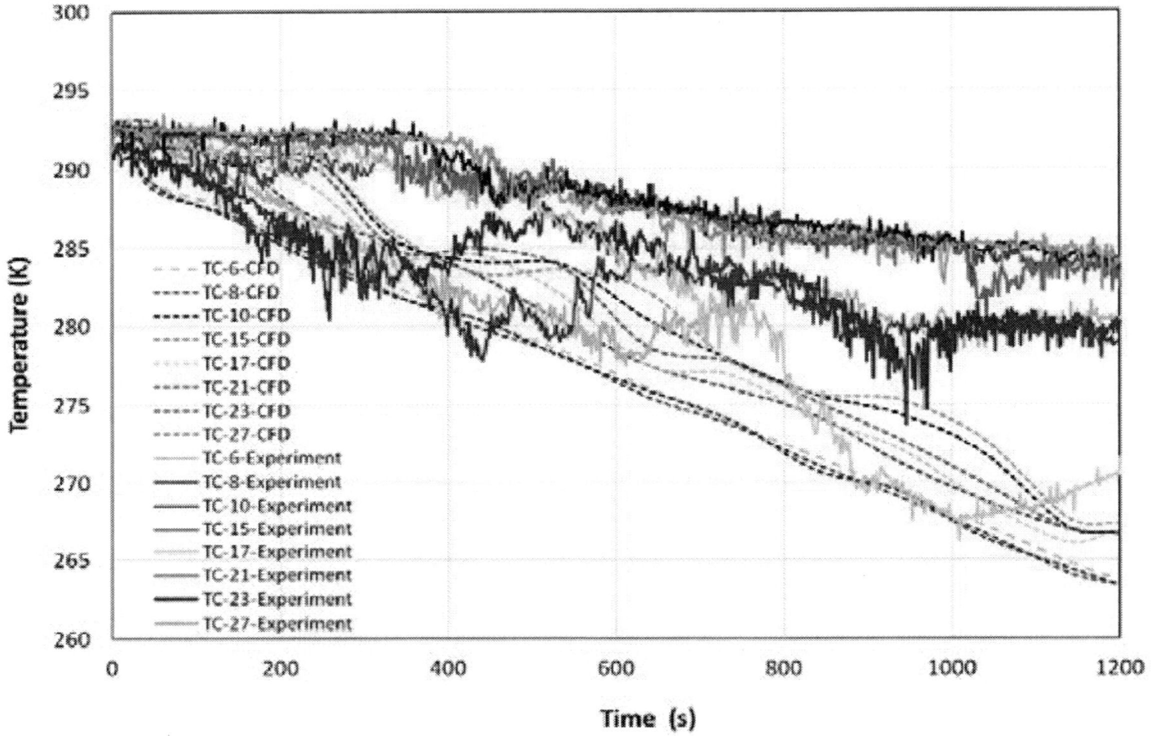

Fig. 7. Water temperature profiles

3.3. Pool Area

A comparison of pool area between MKOPSC trench experiment and LSPREAD simulation is provided in Fig.8. The spreading of LNG pool on water with subsequent water temperature reduction from CFD is also expressed as contours in Fig 9(a) to (d). The contours are obtained from LSPREAD simulation. As the pool spreads, the pool covers the first leg of the trench (t~130s). The CFD results predict the pool area very well up to this duration. As the pool spreads, the pool area reaches a maximum value when the LNG pool contacts the end of leg 2 of the trench (t~400s). The CFD results over-predict the pool area up to this duration. A maximum pool area is maintained for a few minutes where addition of LNG into the pool increases the height, rather than the area. At this stage, the mass inflow to the pool is equalized by the LNG vaporized. It was also observed that due to the high wind speed blowing in the opposite direction of pool spreading in leg 2, rapid vaporization was found to take place instead of pool spreading. As a result of this, the pool was vaporizing steadily in experiment and CFD during the interval 400-700s. Once the pool touches the end of leg 2, it starts regressing. During the time duration of 900-1200s, the pool was spreading and regressing constantly. The time at which regression and pool spreading occurred in the experiment during the time period 900-1200s was not matched well with the CFD. As a result of this, a difference is observed between experiment and CFD results. With further vaporization, the pool area starts reducing, from the end of the second leg of trench. The relative error between LSPREAD and MKOPSC trench experiment was determined to be 21% for pool area.

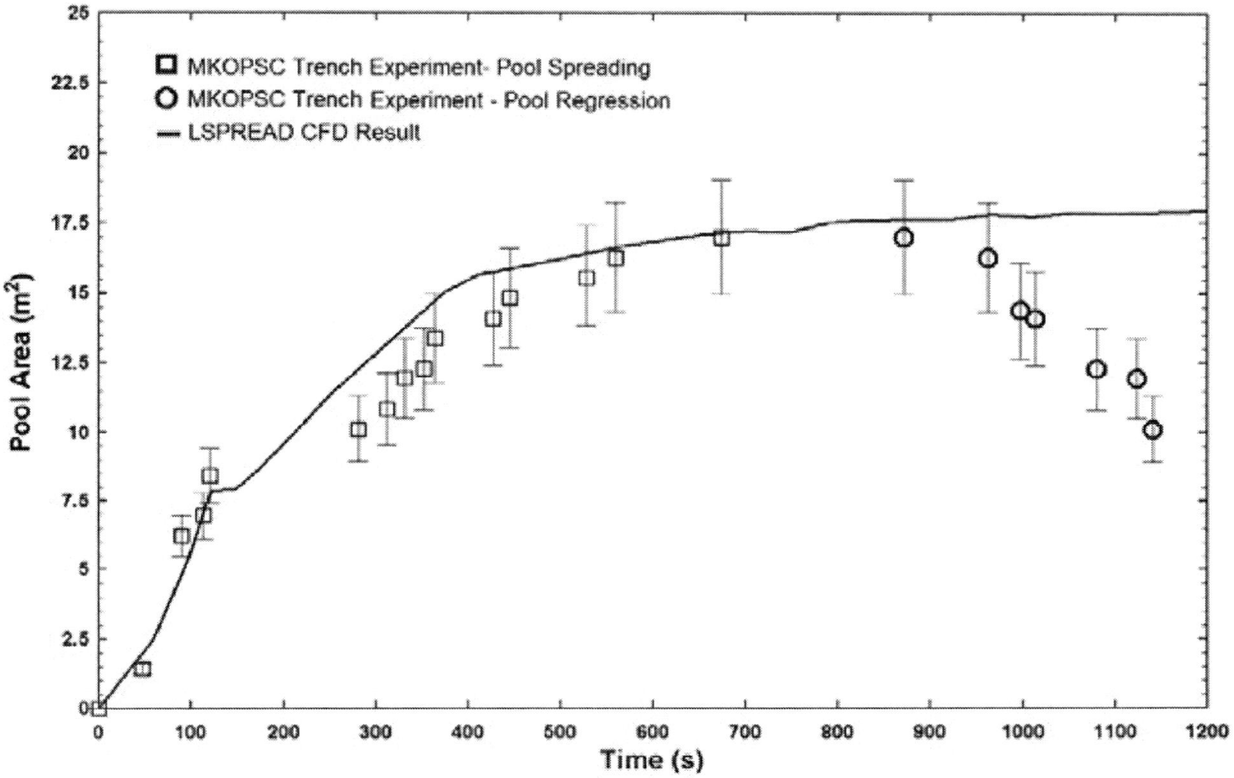

Fig.8. Comparison of pool area between MKOPSC Trench experiment and LSPREAD simulation

3.4. Spreading Rate

A comparison of spreading rate from experiment and simulation is provided in Fig.10. The spreading rate increases initially up to a value of 0.06 m/s till 130s after which, the velocity starts to reduce to a minimum value of 0.02 m/s. However, in CFD, the maximum spreading rate in leg 1 was around 0.08 m/s. During the experiment, wind entrainment was from all direction. However, wind entrainment was restricted to pre-dominant wind direction in CFD. After 130s the pool reaches the boundary of leg 1 of the trench, where it takes a turn to leg 2 causing a reduction in speed. The wall present at end of leg 1 of the trench exerts a drag force on the moving pool that reduces the momentum of the pool. As a result of this, a marked difference in spreading rate is observed between leg 1 and leg 2 of the trench. Fig.10 also shows the velocity contour of LNG from LSPREAD simulation near the discharge area. The momentum of LNG pool near the discharge is very high. As LNG moves away from the source, the momentum of the LNG pool reduces and becomes stable. The behavior of spreading rate between experiment and LSPREAD simulation was similar, however, LSPREAD was found to over-predict the velocity in leg 1. The relative error between LSPREAD and MKOPSC Trench experiment was determined to be 22% for spreading rate.

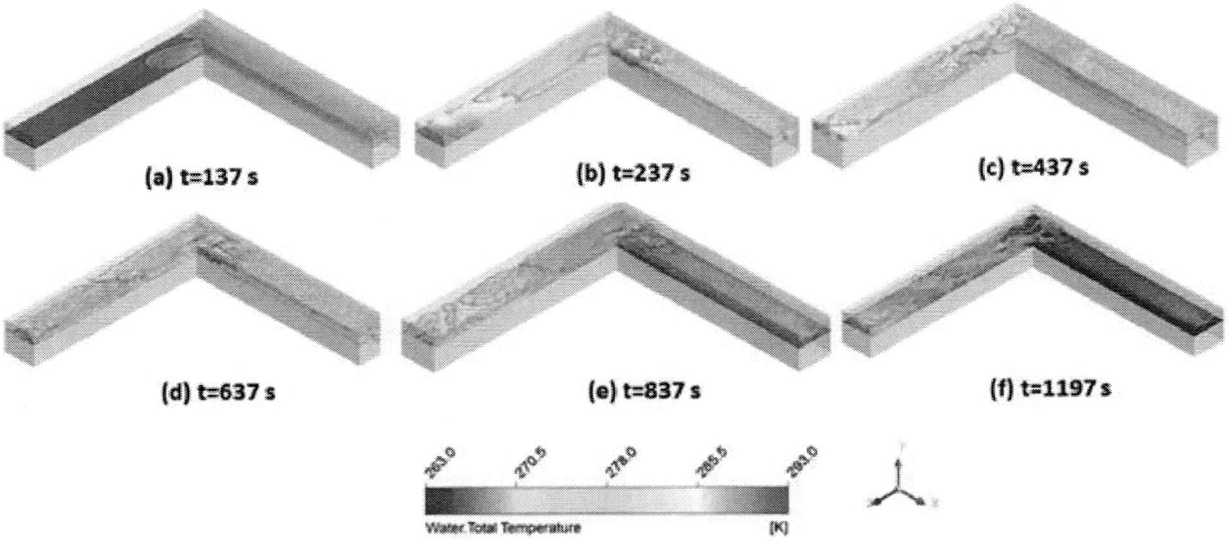

Fig.9. Timed snapshots of water temperature at different time intervals

3.5. Pool Height

Fig.11 provides a comparison of pool height between different experimental measurements and LSPREAD. A difference in various measurements of pool height is observed due to the different location in which it was measured. From Fig.11, we can also observe that the height of the pool tends to increase only after duration of 400s. This is the time taken by the pool to reach the end of leg 2 of the trench. However, the pool height starts to increase earlier in CFD due to higher LNG spreading rate in CFD. The pool height predicted by simulation was representative of the values recorded by manual measurement and array of thermocouples. The ultrasonic level sensor showed the dynamic increase and decrease in pool height due to pool spreading and vaporization. The average relative error between MKOPSC Trench experiment and LSPREAD was found to be 25% for pool height.

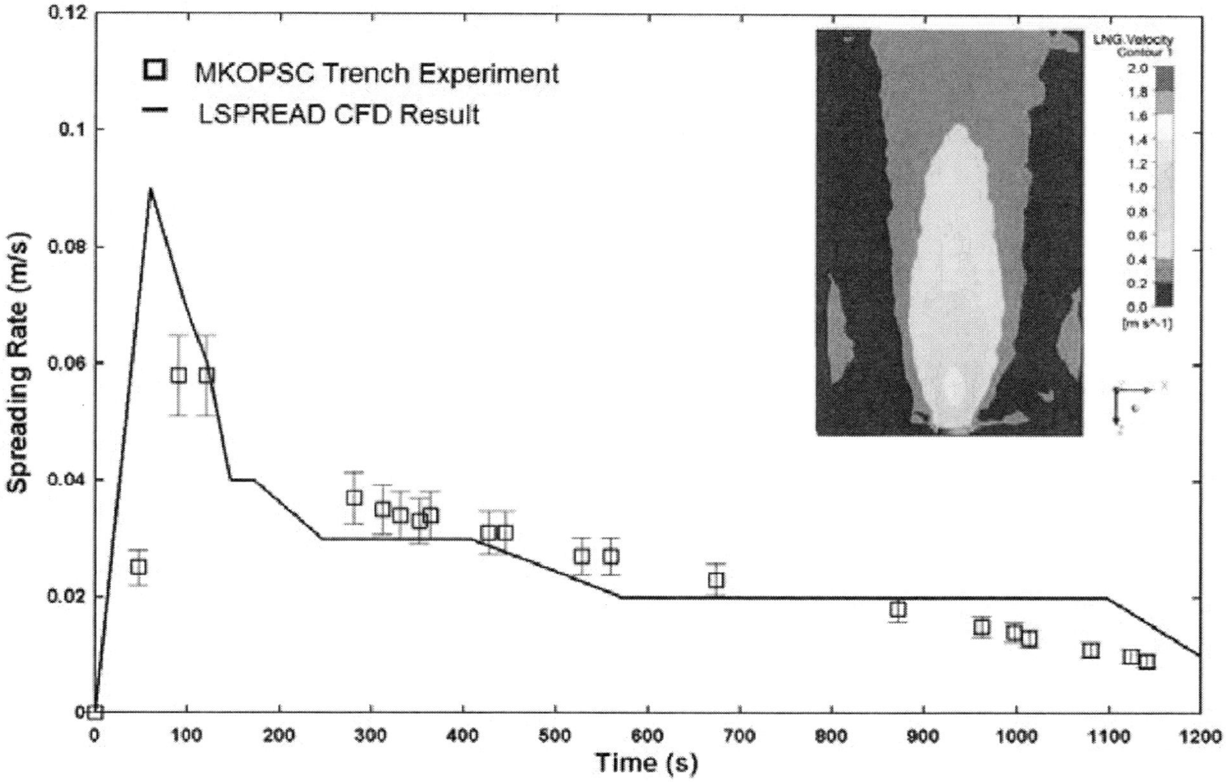

Fig. 10. (a) Comparison of spreading rate between MKOPSC Trench experiment and CFD simulation. Contour from LSPREAD simulation shows the velocity near the discharge area

3.6. Vaporization

Fig.12 provides a comparison of vaporization mass flux between LNG experiments and LSPREAD. The average vaporization mass flux determined from experiment is found to be around 0.2 kg/(m^2s) in the experiment. The vaporization mass flux includes heat input from wind and water. The wind had a significant effect on vaporization in two different ways. Wind provides heat to LNG in the form of convection. Additionally, wind also provides advection in the form of entrainment. The advection process makes the air above LNG unsaturated, which leads to an increase in vaporization. The advection process dominates over the convection when high wind speeds are observed. However, it was difficult to measure the level of entrainment in both experiment and CFD. The high initial vaporization mass flux was due to large temperature difference that is present during the initial stages of the continuous spill. Once the pool covered the entire water surface, the vaporization mass flux was found to remain constant at 0.195 kg/(m^2s) after a time period of 500s. The shift from 0.2 to 0.195 kg/(m^2s) shows a minor change in vaporization mass flux, after 500s. During the time period 500-800s, the pool area was maintained constant. The vaporization mass flux further reduces to 0.19 kg/(m^2s) once the pool starts regressing. The overall relative error between LSPREAD and experiment was determined to be 4% for vaporization mass flux. The values of the current tests are compared with similar LNG experiments. The vaporization rates obtained in this test were comparable to values obtained in Esso Test [8] and Bureau of Mines [7], but over-predicted when compared to Maplin Sands [27] and Shell Test [6].

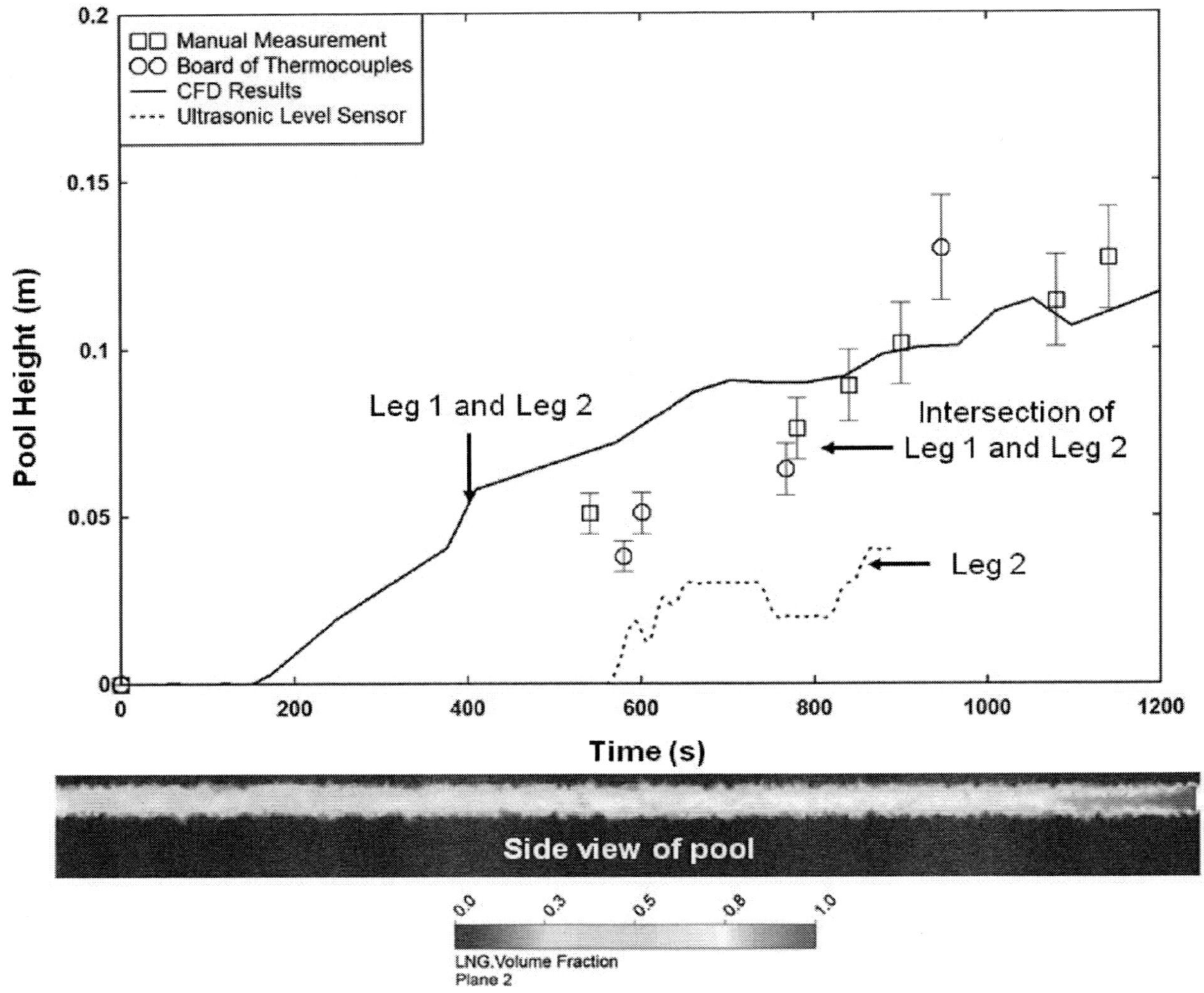

c

Fig.11. Comparison of pool height between MKOPSC Trench experiment and LSPREAD simulation. Contour from LSPREAD simulation shows pool height in leg 1.

4. Conclusion

In this study, a comprehensive experimental and numerical study of LNG pool spreading and vaporization was performed. Measurements of pool spreading and vaporization parameters were made using novel methods. A transient three-dimensional homogeneous multiphase model called LSPREAD was developed in CFX for LNG pool spreading and vaporization on water. A user defined routine for vaporization was implemented in LSPREAD with adaptive time-stepping technique. While modeling, it was noted that there is a lack of information regarding LNG droplet sizes that are formed during the initial stages of release. Further research is required on

LNG discharge modeling to identify the range of droplet sizes based on the LNG storage conditions. The results from CFD were validated with an experiment to simulate the release of LNG on water during side-by side loading operations. Photographic images reveal distinct difference between flashing and vapor cloud formation. Validation results reveal that LSPREAD methodology is capable of emulating the typical behavior of LNG movement in narrow pathways. However, differences were observed between LSPREAD and experimental values due to the ways of determination of source term parameters. It was also observed that wind influenced both pool spreading and vaporization phenomenon. The wind blowing in the opposite direction of the pool tends to impede the spreading phenomenon significantly. The wind also affects the vaporization by providing additional heat and unsaturation through entrainment.

5. Acknowledgements

The authors like to acknowledge financial support from Mary Kay O'Connor Process Safety Center and National Grid for this experiment. The authors would like to acknowledge the help extended by Mr. Kirk Richardson (BFTF) and his team of firefighters for the LNG experiments.

6. References

[1] ABS, "LNG Bunkering : Technical and Operational Advisory," Houston, 2014.

[2] A. Luketa-Hanlin, "A review of large-scale LNG spills: experiments and modeling.," *J. Hazard. Mater.*, vol. 132, no. 2–3, pp. 119–40, May 2006.

[3] D. M. Webber, S. F. Jagger, S. E. Gant, and M. J. Ivings, "LNG Source Term Models for Hazard Analysis : a Review of the State-of-the-Art and an Approach to Model Assessment," Derbyshire, UK, 2009.

[4] H. R. Chang, "Boiling and Spreading of liquid nitrogen and liquid methane on water," *Int. Commun. Heat Mass Transf.*, vol. 10, pp. 253–263, 1983.

[5] G. F. Feldbauer, J. J. Heigl, W. McQueen, R. H. Whipp, and W.G. May, "Spills of LNG on water – vaporization and downwind drift of combustible mixtures," API Report EE61E-72, 1972.

[6] G. . Boyle and A. Kneebone, "Laboratory investigations into the characteristics of LNG spills on water. Evaporation, spreading and vapor dispersion," Chester, Report Number 6Z32, 1973.

[7] D. Burgess, J. Murphy, and M. Zabetakis, "Hazards of LNG spillage in marine transportation," Pittsburgh, Pennsylvania, S-4105, 1970.

[8] W. . May, W. McQueen, and R. . Whipp, "Spills of LNG on Water," in *Operating Section proceedings*, 1973, p. D-140-D-150.

[9] D. M. Webber, "On models of spreading pools," *J. Loss Prev. Process Ind.*, pp. 3–6, May 2012.

[10] F. Briscoe and P. Shaw, "Spread and evaporation of liquid," *Prog. Energy Combust. Sci.*, vol. 6, no. 1, pp. 127–140, 1980.

[11] DNV (Det Norske Veritas), "PHAST consequence modelling software version 6.7." DNV Corporate Headquarters, Hovik ,Norway, 2012.

[12] H. W. M. Witlox, M. Harper, and R. Pitblado, "Validation of PHAST Dispersion Model as Required for USA LNG Siting Applications," *Chem. Eng. Trans.*, vol. 31, pp. 49–54, 2013.

[13] D. M. Webber and M. J. Ivings, "Modelling bund overtopping using Shallow Water

Theory," *J. Loss Prev. Process Ind.*, vol. 23, no. 5, pp. 662–667, Sep. 2010.
[14] N. Gopalaswami, D. M. Laboureur, R. A. Mentzer, and M. S. Mannan, "Quantification of turbulence in cryogenic liquid using high speed flow visualization," *J. Loss Prev. Process Ind.*, vol. 42, pp. 70–81, Jul. 2016.
[15] S. J. Kline and F. A. McClintock, "Describing Uncertainties in Single-Sample Experiments," *Mech. Eng.*, vol. 75, pp. 3–8, 1953.
[16] R. Marcer, B. Yerly, L. Pomié, B. Lequime, M. Rivot, E. De Carvalho, and F. Baillou, "CFD Simulation of LNG spillage," in *6th European Conference on Computational Fluid Dynamics(ECFD VI)*, 2014, p. 12.
[17] T. A. Cavanaugh, J. H. Siegell, and K. W. Steinberg, "Simulation of vapor emissions from liquid spills," *J. Hazard. Mater.*, vol. 38, no. 1, pp. 41–63, Jul. 1994.
[18] E. M. Drake, A. a. Jeje, and R. C. Reid, "Transient boiling of liquefied cryogens on a water surface I. Nitrogen, Methane and Ethane," *Int. J. Heat Mass Transf.*, vol. 18, no. 12, pp. 1361–1368, Dec. 1975.
[19] E. M. Drake, A. A. Jeje, and R. C. Reid, "Transient Boiling of Liquefied Cryogens on a water surface. II Light Hydrocarbon Mixtures," *Int. J. Heat Mass Transf.*, vol. 18, no. 12, pp. 1369–1375, 1975.
[20] ANSYS Inc, "Ansys CFX: Release 15" Ansys Inc, Canonsburg, PA, 2016.
[21] B. E. Launder and D. Spalding, "The numerical computation of turbulent flows," *Comput. Methods Appl. Mech. Eng.*, vol. 3, no. 2, pp. 169–289, 1974.[22] N. Gopalaswami, R. A. Mentzer, and S. Mannan, "Investigation of pool spreading and vaporization behavior in medium-scale LNG tests," *J. Loss Prev. Process Ind.*, vol. 35, pp. 267–276, 2015.
[23] P. J. Berenson, "Film-Boiling Heat Transfer From a Horizontal Surface," *J. Heat Transfer*, vol. 83, no. 3, pp. 351–356, 1961.
[24] R. Qi, D. Ng, B. R. Cormier, and M. S. Mannan, "Numerical simulations of LNG vapor dispersion in Brayton Fire Training Field tests with ANSYS CFX.," *J. Hazard. Mater.*, vol. 183, no. 1–3, pp. 51–61, Nov. 2010.
[25] N. Gopalaswami, T. Olewski, L. N. Véchot, and M. S. Mannan, "Small-scale experimental study of vaporization flux of liquid nitrogen released on water," *J. Hazard. Mater.*, vol. 297, pp. 8–16, Oct. 2015.
[26] S. T. Chan, "Numerical simulations of LNG vapor dispersion from a fenced storage area," *J. Hazard. Mater.*, vol. 30, no. 2, pp. 195–224, Apr. 1992.
[27] J. S. Puttock, D. R. Blackmore, and G. W. Colenbrander, "Field experiments on dense gas dispersion," *J. Hazard. Mater.*, vol. 6, pp. 13–41, 1982.
[28] J. Sienfield and S. Pandis, *Atmospheric Chemistry and Physics: From Air Pollution to Climate Change*, 2nd ed. new Jersey: John Wiley & Sons, Inc., 2006.

Nomenclature

A	Interfacial area density (1/m)
C_p	Specific heat capacity (J/kg K)
D	Droplet diameter (m)
E	Internal energy (J/kg)
f	Pool spreading or vaporization parameter

g	Acceleration due to gravity (m/s^2)
h	Heat transfer coefficient ($W/(m^2K)$)
H_v	Latent heat of vaporization (J/kg)
k	Thermal conductivity (W/(mK))
L	Characteristic length of the pool (m)
n	Thermocouple tolerance value (K)
m	Mass flux between phases ($kg/(m^2s)$
Nu	Nusselt number
p	Pressure (Pa)
Pr	Prandtl number
Re	Reynolds number
r	Volume fraction
q	Heat flux (W/m^2)
$\dot{Q}$	Heat Source (W/ m^3)
t	Time (s)
T	Temperature (K)
U	Velocity (m/s)
x,y,z	Cartesian coordinates

Greek Letters

μ	Viscosity (kg/ms)
ρ	Density (kg/m^3)
κ	Turbulent Kinetic Energy (m^2/s^2)
ε	Eddy Dissipation Rate (m^2/s^3)
σ	Surface Tension (N/m)
τ	Stress Tensor

Subscripts

air	Air
d	Data retrieval time
e	Experimentally determined value
LNG	LNG liquid phase
lam	Laminar regime
n	Tolerance of thermocouple
r	Response time
v	LNG vapor
water	Water

513358 How to Partner up and Protect Your Intellectual Property in Energy Industry Joint Ventures

Abstract:

Exploration and development of oil and gas reserves increasingly requires costly, tailored technologies and enhanced safety and environmental measures. This is due in part to greater desire to develop unconventional oil and gas resources, such as shale gas and tight oil, and to develop reserves in more remote and environmentally-sensitive areas. Exploration and production companies have been developing deepwater reserves, shale plays, and other unconventional sources, but many lack the infrastructure to gather and process the resulting oil and gas. As a result, the industry has seen a rise in joint ventures between upstream and midstream companies that have the capacity to transport and store gas or crude oil, as well as joint ventures between two or more upstream companies. This includes both agreements between domestic companies and international joint ventures.

Although joint ventures can help companies spread capital costs, overcome technological limitations, and mitigate risks, they also create potential pitfalls for companies with intellectual property assets. For example, a company may need to grant its joint venture partners a license to use its patented technologies or even share its trade secrets and other know-how with those partners, who may be industry competitors. Additionally, joint ventures can lead to development of new innovations, and disputes may arise as to who owns the resulting intellectual property and has the right to use it. Protecting these valuable intellectual property assets is an aspect of managing joint venture and joint development projects that is often overlooked. This presentation will outline potential perils of joint ventures in the energy industry and discuss strategies to protect intellectual property portfolios when entering these agreements.

Project Changes in Gas Monetization: Pitfalls or Successes

Jeffery Gaspard, Ph.D., P.E.
KBR Process Engineer
AIChE– April 24th, 2018

18AIChE Spring Meeting

Looking at potential Gas Monetization projects can be a daunting task. What can make them more daunting is the prospect of changes in the development and operation of the project. Some changes can be managed during the life of the project: some raw material sources, contract price for finished goods, etc. But other changes can drastically change the project; some move the project from viable to uneconomical. Most of these changes are out of the control of the operator, but with proper planning, some changes can be mitigated during the design phase of a project. This presentation will talk about some considerations that need to be discussed and planned during the early stages of a project in order to maximize the success of the project.

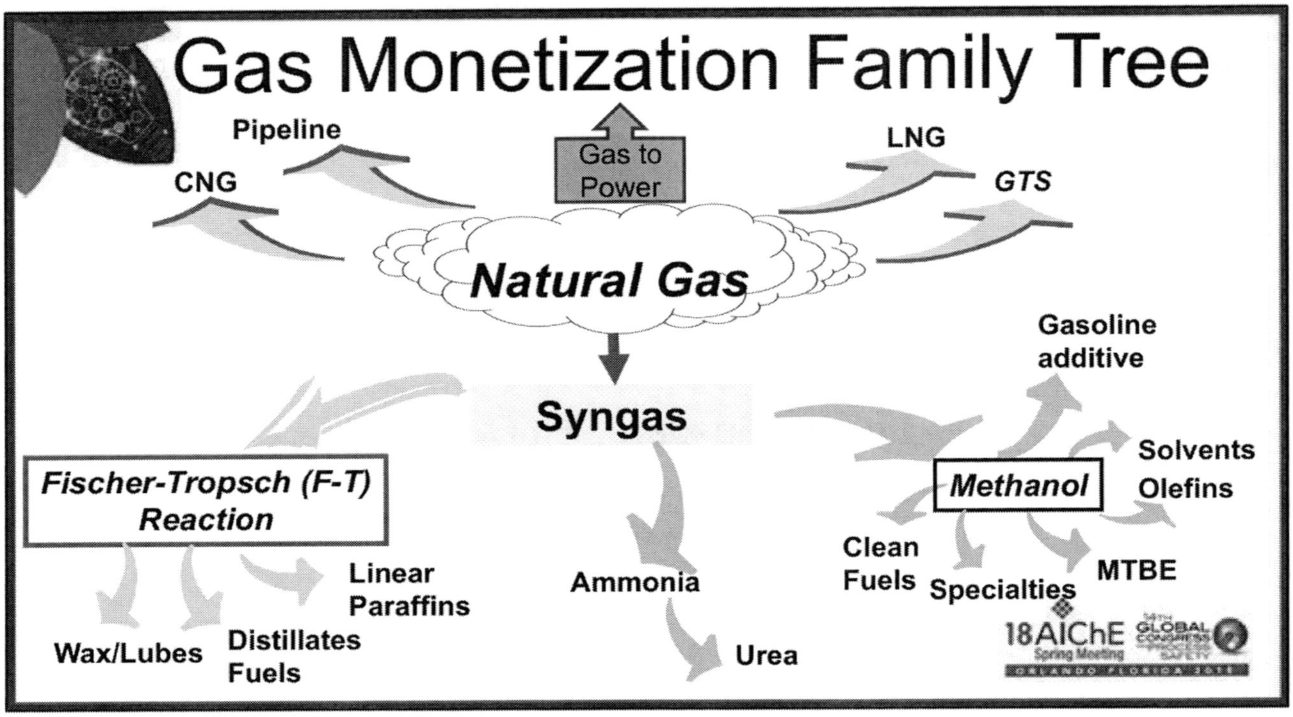

The overall "family tree" of the possibilities of Gas Monetization. Many products and services can be rendered from Natural Gas. This presentation will cover ideas that are applicable in almost all of these markets.

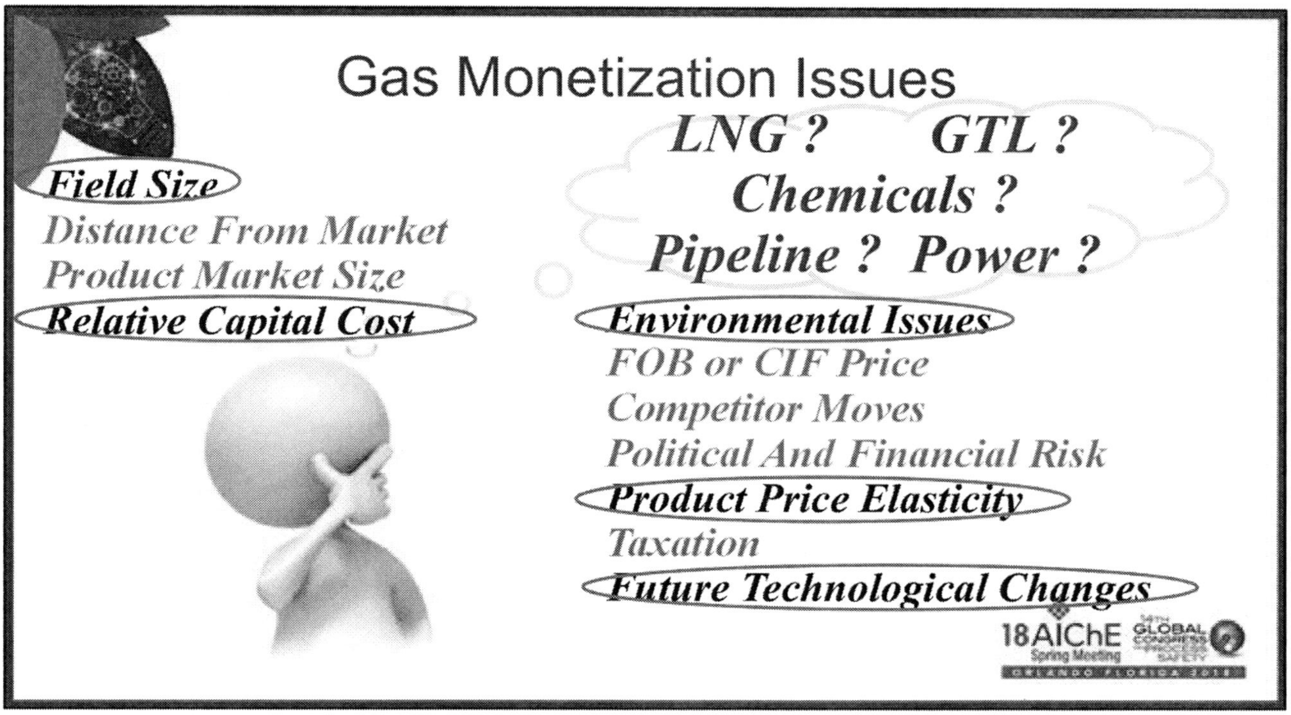

While there are many factors that play into a projects profitability or viability, these highlighted will be mentioned or discussed in this presentation.

- Field size plays off of your production rate and can impact shipping to the market
- Relative Capital Cost will be discussed in shipping options. Obviously changes in shipping methods changes costs, but small increases in cost can be beneficial to short term profitability during disruptions
- Environmental issues are important in any project. The whole life of the project should be considered for changes in environmental impacts. What is OK today might not be good enough in the future, even with grandfathering.
- Product price elasticity is discussed with shipping and input changes. During disruptive events, being flexible means you can command a premium for your 'normal' product.
- Future technological changes will be discussed in two topics: delivery and feed gas (input) challenges. Competitors can react to market changes, especially in feed (input) cases faster than an existing plant. How is your design capable of process changes, if necessary?

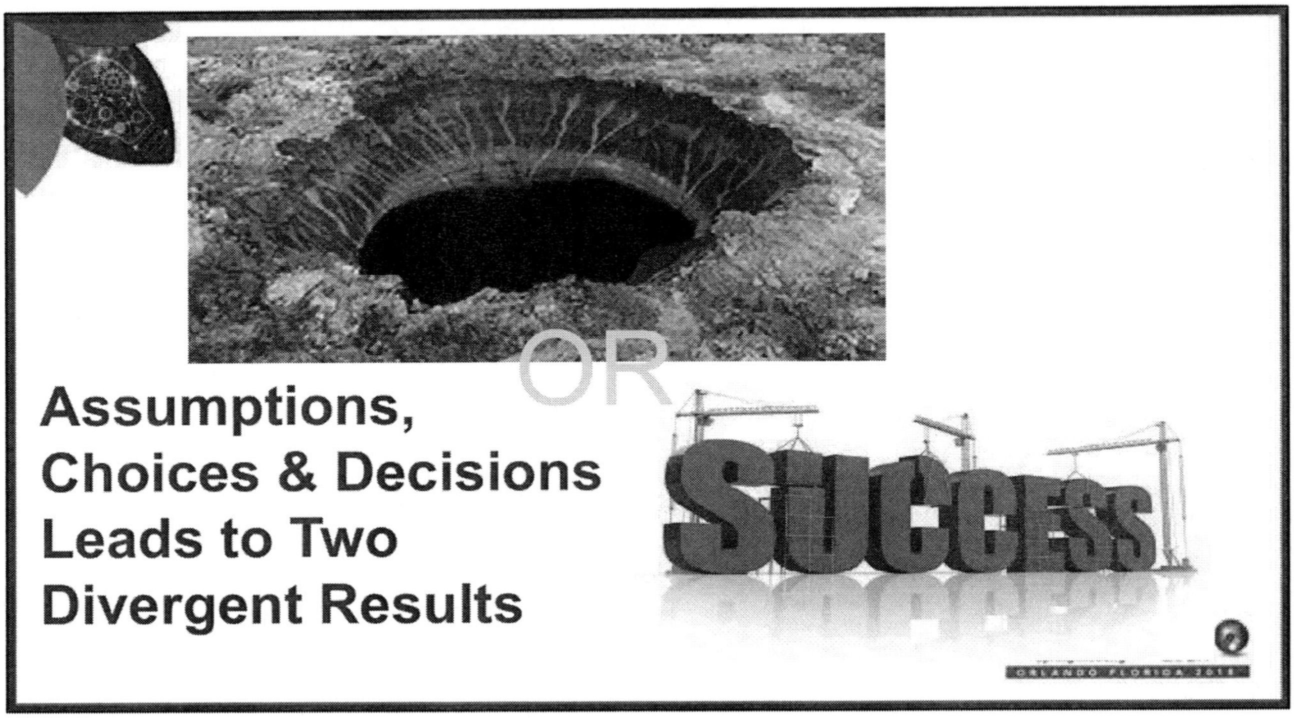

With these topics in mind, let's talk about three specific topics that can lead a project to being a pitfall or leading to success.

Basic Assumptions, Choices & Decisions Leads to Two Divergent Results.

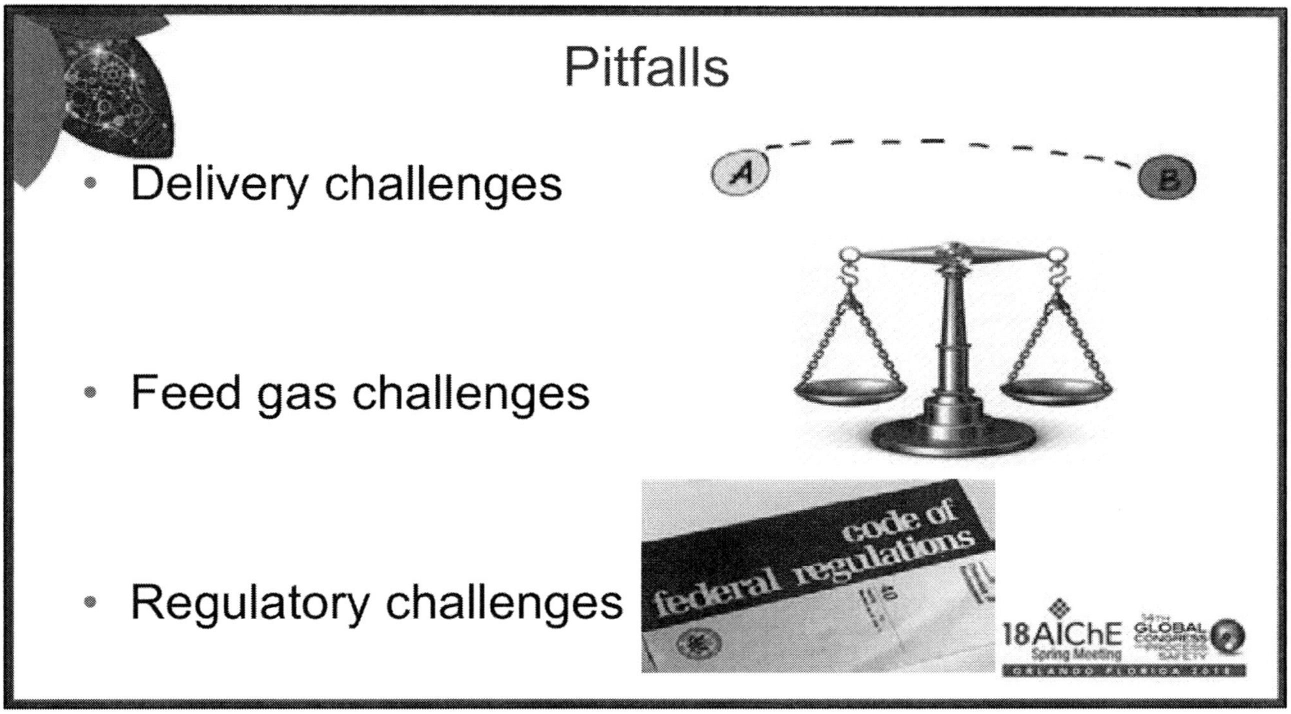

The three main topics for today's presentation are:

- Delivery Challenges
- Feed Gas Challenges
- Regulatory Challenges

Each of these topics can impart success or pitfalls on any Gas Monetization project. Each one will be covered during this presentation.

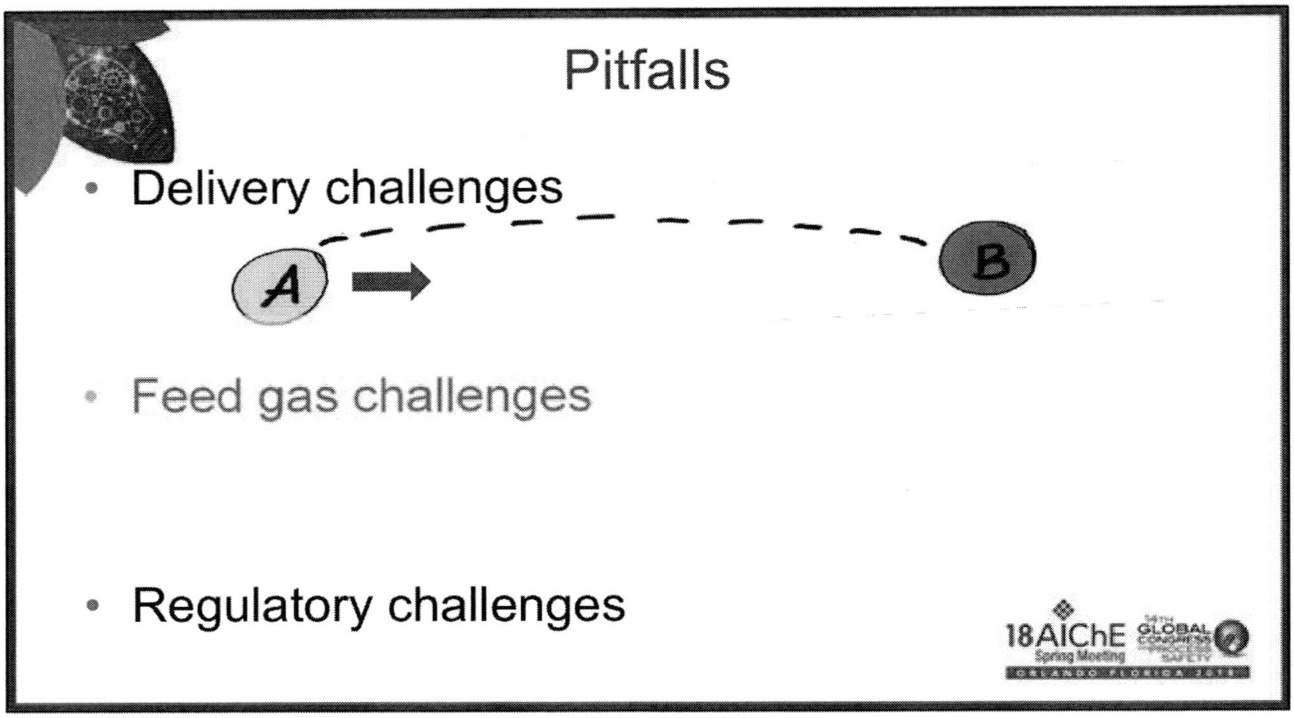

The first topic is Delivery challenges for Gas Monetization

Delivery Challenges

- Weather
  - Flooding, hurricanes, power outages
- Shipping outages
  - Rails, roads, obstructions to waterways
- Site Storage
  - Stop a running plant

Once a product is made at a Gas Monetization facility, it needs to get to the market or user. During normal operations and conditions, this can be taken for granted. But during disruptive times, small changes could have a great impact on the facility.

Things that can cause disruptive times are weather based (flooding, hurricanes (leading to flooding)), economic outages (power outages, road/rail disruption), political/economic unrest (workers, lack of gov't) or other basic things as site storage (tank failure/damage, turn-around schedule). Other challenges can be input related; a pipeline reduction or restriction can limit overall production or slow utilities which hinder production. While generally we think of delivery as a plant output issue, it can be a plant input issue as well. One example from my work experience was a lack of cardboard tubes for a polymer plant. Without them, total production shutdown.

To look at potential solutions for delivery challenges, we need to know how is the product moved to the market. If the product is shipped out via rails, is there a possibility for trucks? If it is shipped in large vessels, can you barge out as well? Is there a potential for pipeline to go with shipping? If pipelined out, can you store for long periods on site until pipeline is restored?

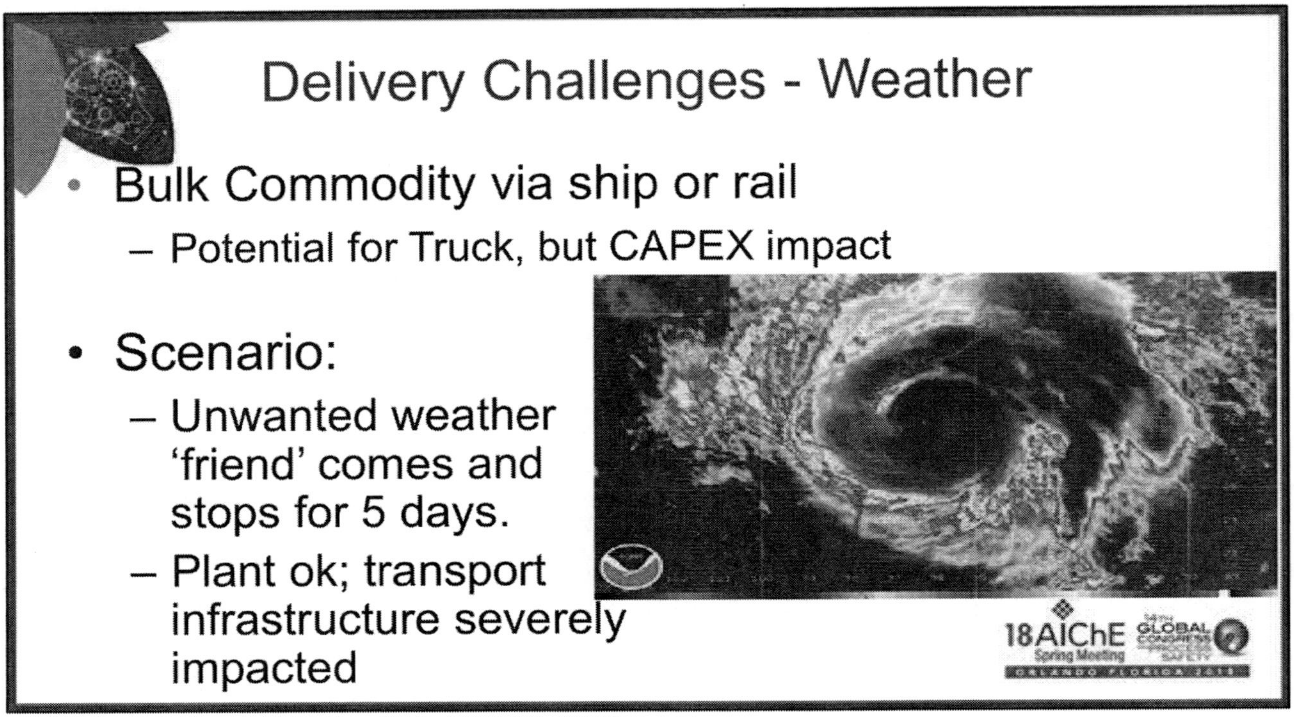

Looking at the delivery challenges, let's look at the impact from weather.

Let's say an unwelcomed friend comes to visit; his name is H. Harvey. Wait, too specific; let's call him Hurricane H. He comes to shore, but doesn't understand he needs to leave. In the wake, he drops about 48 to 65 inches of rain in 4 to 5 days (122 to 165 cm of rain). Roads, rail, ports, airports, etc. are all shut down. Some roads are damaged and until repaired, unusable. Rail bridges over rivers and some basic crossing are out. What do you do? Can your plant handle production once you get back up (after shutting down for Hurricane H)? Will you have to stay down for an extended period due to this event? Northerly customers, who weren't affected by the storm, can still receive products and have limit stock on hand. If they shut down due to lack of product, they may not come back to you or look elsewhere for raw materials/their process input.

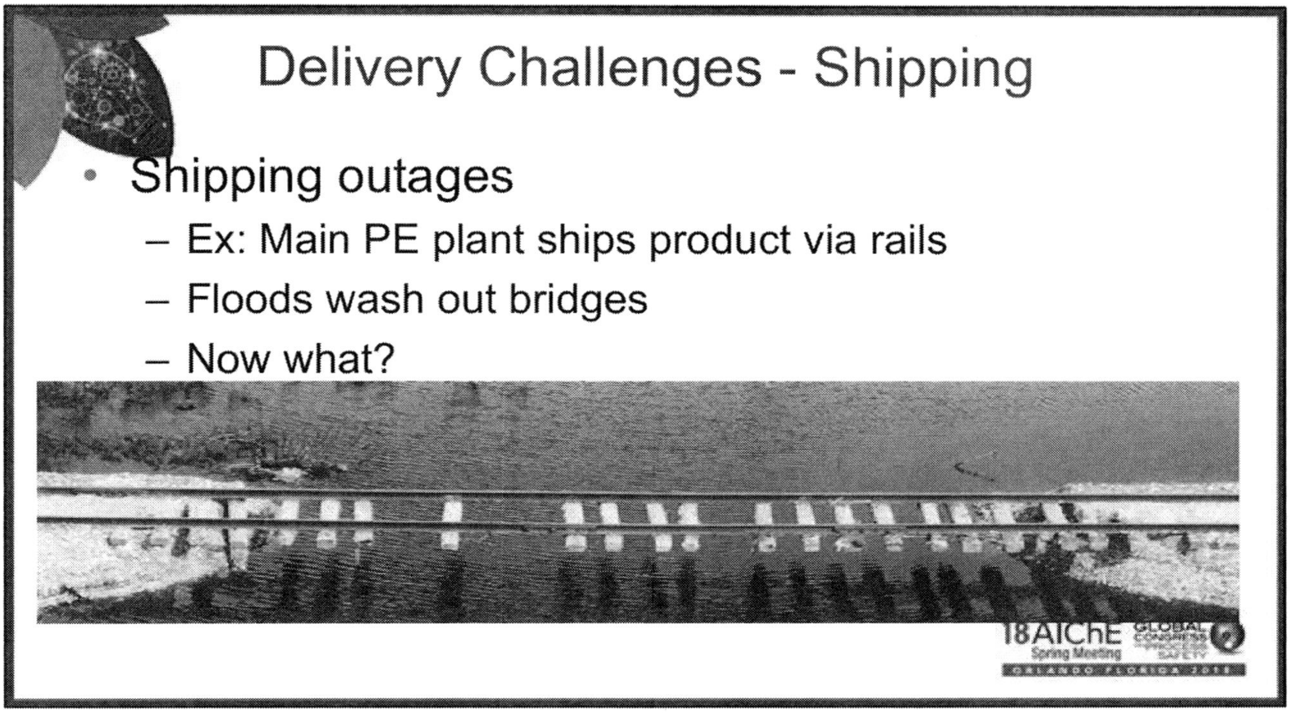

Your plant, that was affected by Hurricane H, uses primarily train loading for product moving. But with train tracks out, you are idle. But, with foresight, you installed a truck loading facility during a turn-around. While it ships less per load, you are able to move product out once most roads are functional. While this is an increased cost for shipping (4 trucks is about 1 railcar) and increased personnel issue (truck leader through plant and security gates), you are up and producing (maybe at full rates, maybe at reduced rates) compared to other competitors in the area. Shipping times are about the same (maybe faster), but your customers can stay up and running with little to no disruption. You might be able to command a premium if your competitor's customers cannot get product and come to you.

Also working with this scenario is looking at any potential product wheel/schedule. You might move higher profitability products or higher rate products to take advantage of the economic situation.

In installing the truck loading, originally thought of as a 'losing item' or 'wasted money', the unit might prove to be the reason you maintain profitability or production goals during disruptive times.

Your plant, that was affected by Hurricane H, uses primarily train loading for product moving. But with train tracks out, you are idle. But, with foresight, you installed a truck loading facility during a turn-around. While it ships less per load, you are able to move product out once most roads are functional. While this is an increased cost for shipping (4 trucks is about 1 railcar) and increased personnel issue (truck leader through plant and security gates), you are up and producing (maybe at full rates, maybe at reduced rates) compared to other competitors in the area. Shipping times are about the same (maybe faster), but your customers can stay up and running with little to no disruption. You might be able to command a premium if your competitor's customers cannot get product and come to you.

Also working with this scenario is looking at any potential product wheel/schedule. You might move higher profitability products or higher rate products to take advantage of the economic situation.

In installing the truck loading, originally thought of as a 'losing item' or 'wasted money', the unit might prove to be the reason you maintain profitability or production goals during disruptive times.

Delivery Challenges - Shipping

- Truck loading allows rapid movement
- Prevents costly storage fees for product
  - Tankers vs. rail cars vs. plant shut down
- What are competitors doing?
  - Geographically similar risks?
  - Encompass all methods?

In looking at any potential new or existing facilities, look at the main method of shipping the product out during normal times. Is there a way you could incorporate an alternative method? While trucking is not the best method for some products, say LNG, it does allow for flexibility of moving within a land-mass. While shipping a 'smaller' rate via trucks is not ideal, you can move a product rapidly. Also, by shipping something out, you can maintain production, albeit a reduced rate. This can prevent a costly or timely shutdown/start-up of a plant (utilities, off-site storage, etc.) or an environmental issue of flaring/thermal destruction of 'good' material.
Another method is on-site storage. Can you utilize rail cars or built/lease temporary storage on unused land? Would increasing on-site storage by 1 day or 1 week 'save' a shut-down? Would the costs even out?
Finally, what is your competition doing? Is everyone looking only at ship loading and ignoring rail or truck shipping? If your main method is pipeline, can you ship via truck or rail in the event of pipeline disruption? In going back to LNG, if you ship load, could you change ship sizes or load shipping containers? Would this open new markets (think LNG for ship fuel or LNG for 18-wheeler fuel) vs. traditional current markets?

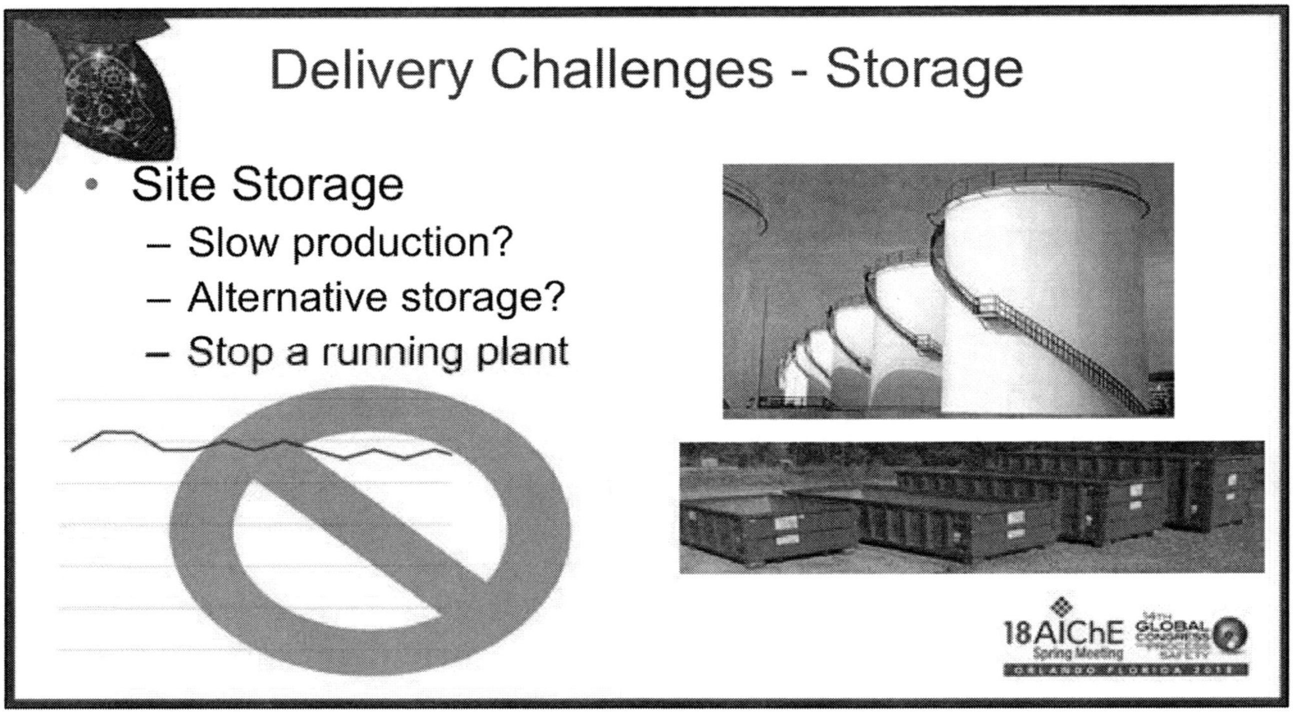

There are other challenges with storage and changes to shipping. If you store everything in 30,000 MT storage tank, it can be hard to get just 2 MT out to fill truck instead of rail/barge. Having smaller storage or alternative loading facilities will allow for nimble shipping and alternative means to sell production. There are utility requirements for tanks and generally larger tankage might be better, but having a mid-sized tank for smaller operations can be beneficial.

With a continued LNG example, by having a smaller tank that is used for railcar or ISO container loading, you can move product out of the site during a disruptive time (contracted buyer has a hurricane or political unrest). Having additional storage may help if you are limited in contractual production, but can produce more. For example, if you are limited to X number of ships per year due to waterway congestion, having a truck loading area would allow for additional production without expensive ship movement/purchasing. Truck/rail loading could lead to other markets, such as 18-wheeler or ship fueling (LNG as fuel).

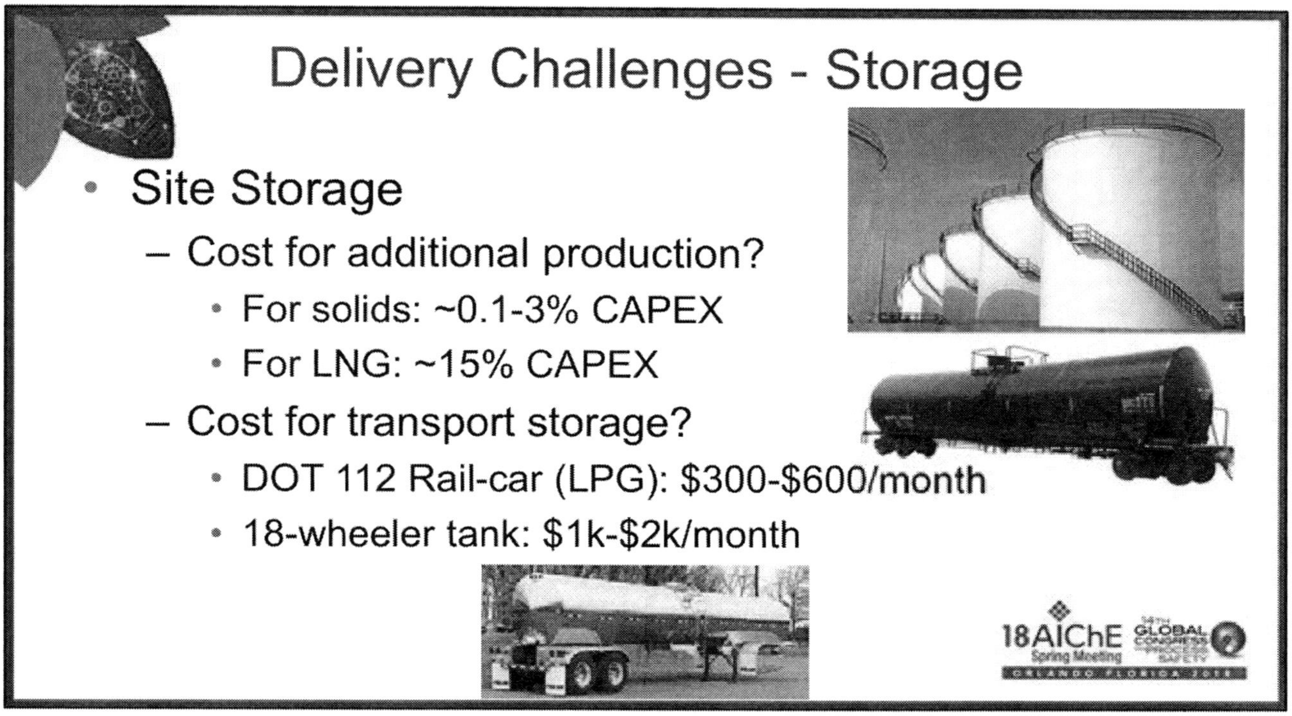

With storage issues comes cost. While storing cryogenic liquids is challenging, storing solids or 'normal' liquids is much easier. In some cases, solids can be placed on prepared, level ground with minimal prep work. Other times, it is a thin-walled tank, mostly to keep the elements out. For 'normal' liquids, such as medium-density hydrocarbons, the tank requirements are higher than solids, but much less than cryogenic or low vapor pressure HC. A great example would be diesel fuel storage. The additional costs for extra storage might be in the less than 1% CAPEX of the project cost.
For other liquids or gases, storage costs can start out steep and get worse. The increased CAPEX cost greatly exceed the OPEX cost of a shut-down/start-up and do not make much economic sense. Only under extreme cases would additional storage help out a project.

Another method of storage is short-term storage. This can be tanker storage or roll-off storage. These costs are more short-term for specific time-frames. Looking at the rail costs, these are for 3-12 month leases; the 18-wheeler tank car is for month-to-month. If short-term storage is used, space is needed to hold the tanks. If rail is used, additional rail lines will be needed. For wheeled tank trailers, empty land is needed. This can be the cheaper option, since storage areas can be leveled ground with crushed rock; this could be a re-purposed lay-down yard.

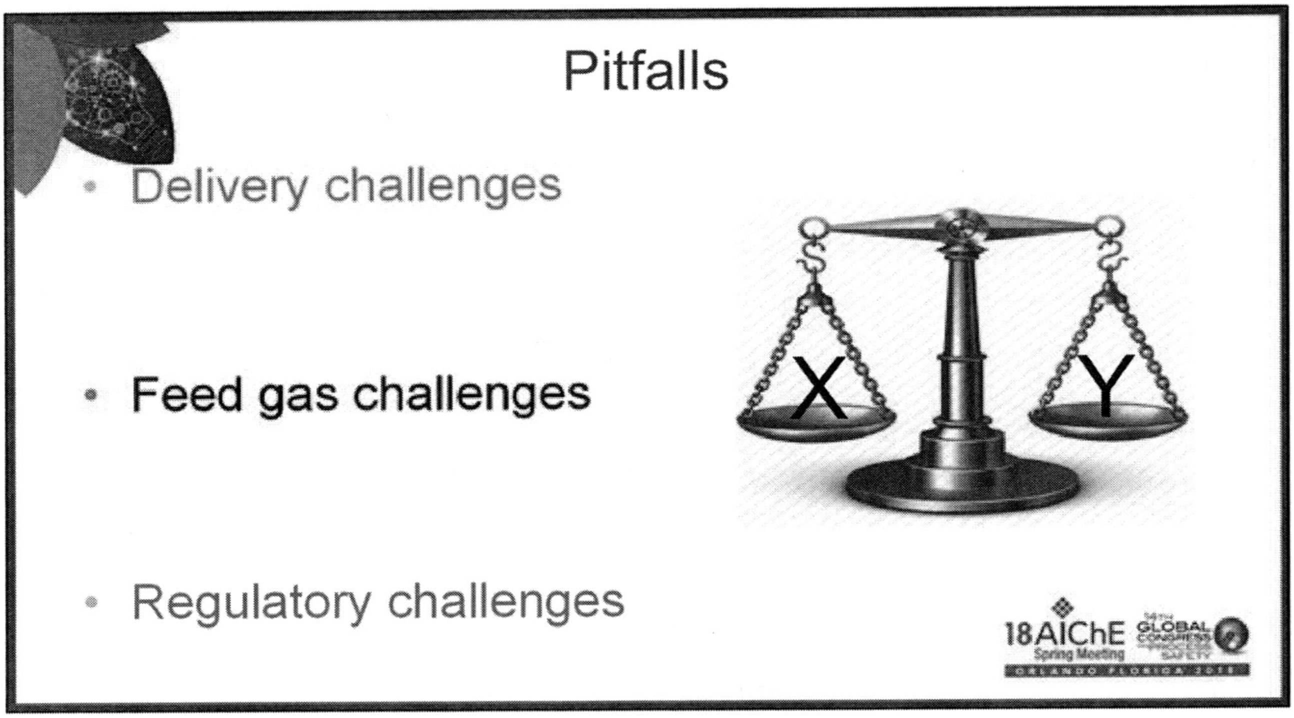

The next topic to be covered is Feed Gas challenges. This will look at composition or other challenges. This can happen most with industries that deal with raw or unfiltered inputs, such as raw gas from wells or comingling of several product streams with broad product specifications. In industries with tighter input requirements, this is less of a concern (LNG vs. polyethylene). The delivery challenge will assume to be fixed since we just finished talking about it.

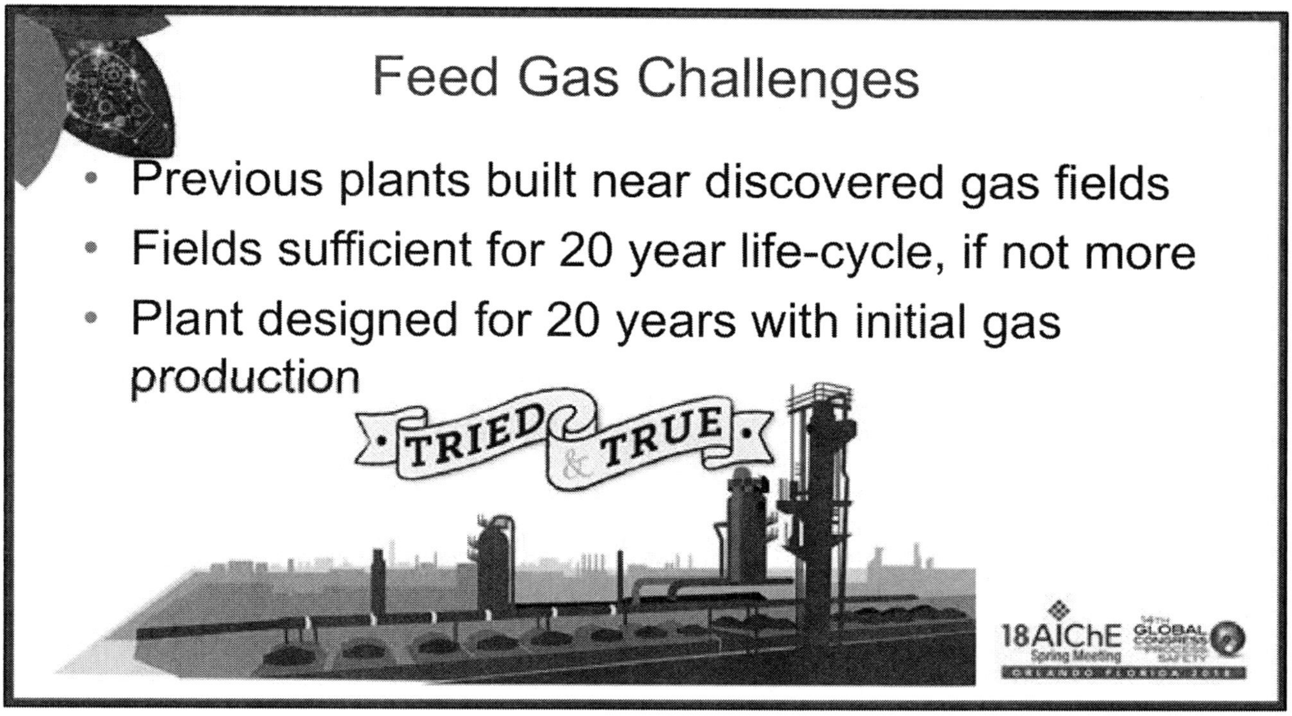

A feed gas challenge is something that, for North American LNG, is something new. Traditionally LNG plants know what the feed gas is, can anticipate the changes over the life of the project/wells and adjust operations accordingly. And generally, for LNG baseload plants, the gas is 'single sourced'; i.e., from a single field (multiple wells, but the same overall field). This is what I would call the "Tried and True" method of design.

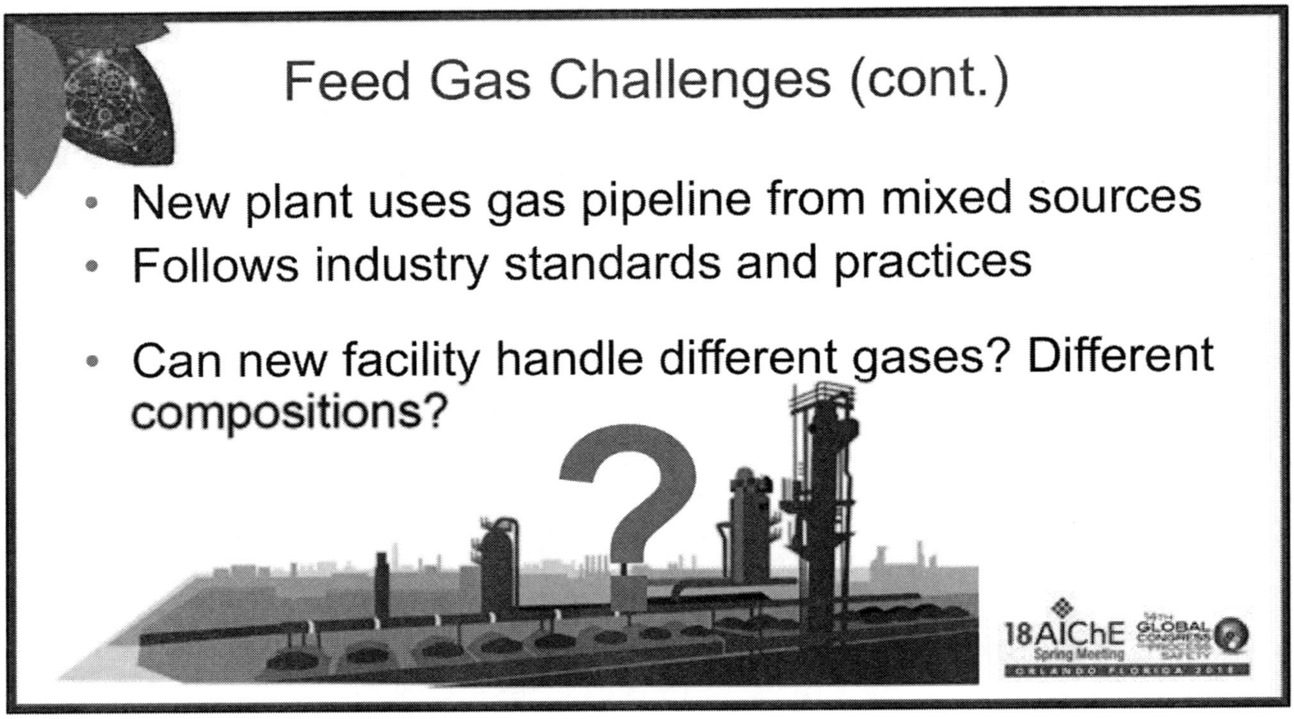

With new LNG projects in N. America, the gas is not the same as before. Using the older method of design is great if you know the gas feed is constant or subject to slow changes. But using pipeline gas, gas that has been cleaned up but follows general guidelines, has a different set of challenges. No longer is the plant fed by one well or field. Several fields can be comingled into one gas feed; there could be a situation where the gas is LNG brought from one location, gasified and put in the pipeline. A N. American LNG plant could take the gas and re-liquefy it. In traditional plants with traditional wells, a gas field will change over time, but usually measured in months/years. With pipeline gas, the gas can stay within specifications of the pipeline, but change composition quickly. This can be measured in days or hours. This presents a much more dynamic operational challenge. This can impact overall production if equipment is not sized correctly.

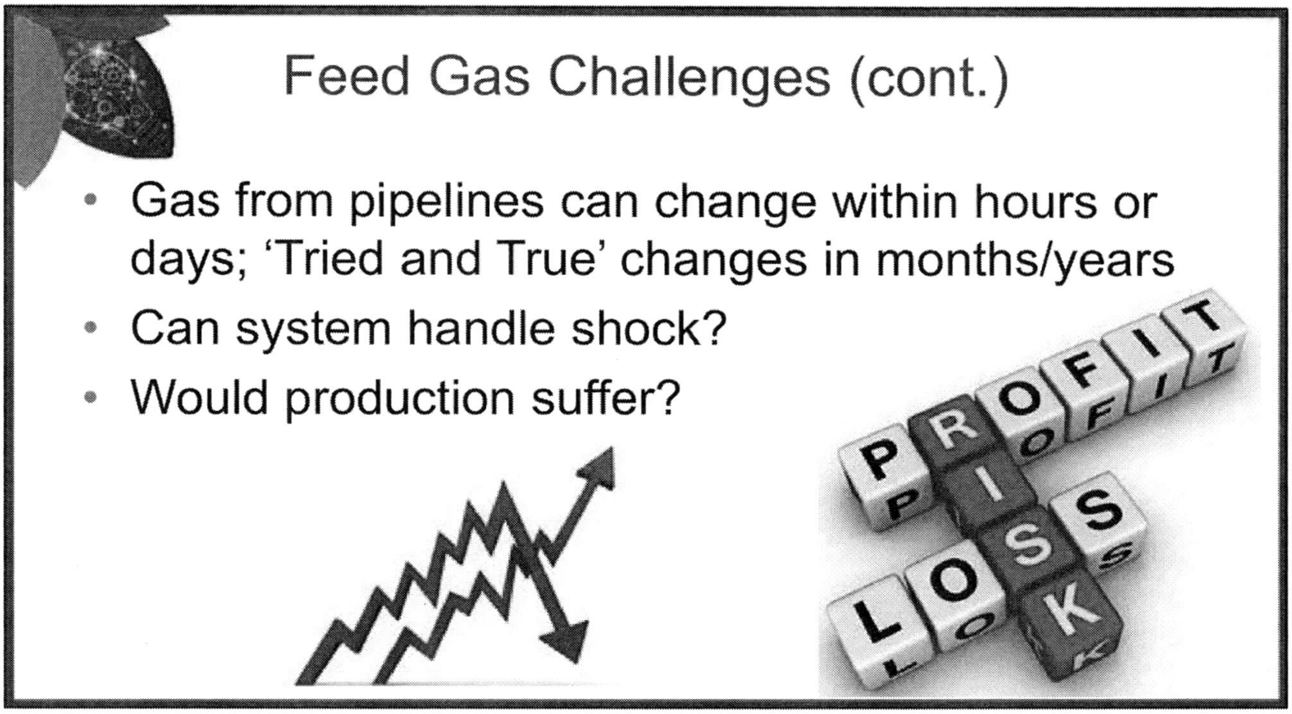

With the traditional LNG plants, the 'Tried and True' design, the gas fields age and change over time. But as the changes are related to the rate of withdrawal, usually by only the LNG plant, these can be monitored and adjusted to enhance production. Additionally, any work done on the gas fields can be relayed to the LNG plant and operations adjusted to upstream gas work.

With newer, pipeline gas LNG plants, the gas comes from many different sources. With different sources of gas, wells and underground storage, along with seasonal demand changes, pipelines can change faster. Work done on wells is not communicated to the LNG plant, usually because the gas will pass through a natural gas processing plant first. But work done to a NGP plant usually will not be communicated to any downstream users, unless major disruption is expected. So knowing what is 'coming down the pipe' is more uncertainty, which leads to increase in risk. With the increase in risk, there will need to be an increase to CAPEX to cover said risks (increased acid gas scrubbers, dehydration, etc.)

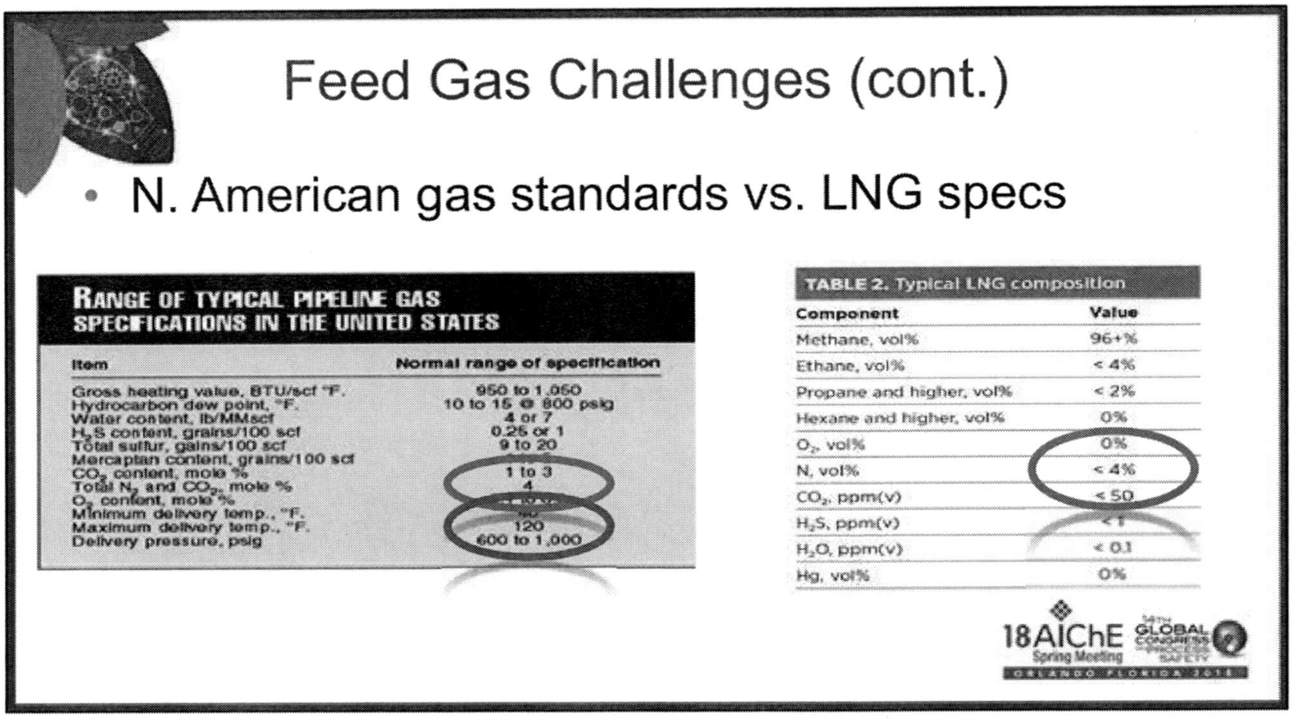

Presented here are 2 tables: one for the typical pipeline gas in North America and a typical LNG spec for production.

We see that pipeline gas is pretty good, compared to some well gas, but no where near the needs, typically, for straight-to-LNG production. There will still be a need for gas clean-up. Only in the fringes do we see or hear about using 'US pipeline gas to LNG straight'. But how much do you need to clean the gas can greatly affect your plant and the CAPEX and OPEX values. CO_2 traditionally has held at a max of 2%; going to 3% would add a significant cost to the project. Adding the extra 1% could mean a separate train for CO_2 clean-up. Instead of a 1-1 ratio for an LNG train/unit count, the Acid Gas clean-up would be a 2-to-1 ratio. Along with changing CO_2, sulfur can be a nuisance to clean up as well. And changing the pressure of delivery can cause major efficiencies problems for the liquefaction section (as well as dehydration units).

A question that has come up is why is the US gas so 'out of sorts' with LNG specs? To fully answer that, we need to look at who the users of N. American gas are and who's driving the overall composition/requirements.

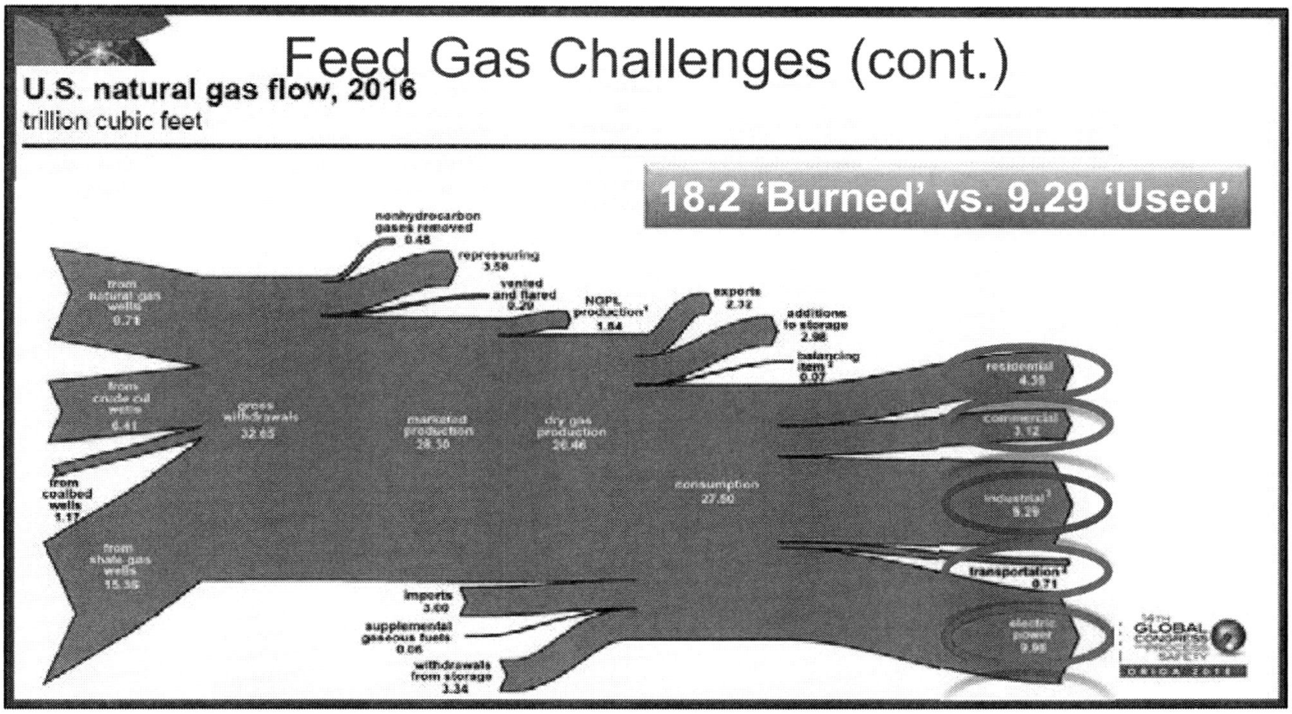

Why is there an issue with changing pipeline gas quality and would it effect other downstream users? To answer easily, no. Why? Most of the downstream users are burning natural gas. As long as it heats (follows the heating value and contaminates), why be concerned if it changed from 1% CO_2 to 3% overnight? Does the extra N_2 in the stream effect the performance of your unit? Power generation might notice a bit, but if the Wobbe index holds, they might not care. So for a majority of the users in North America, gas specifications is something that is not much of a concern. And until a larger share of the gas use is for LNG (either buried in the exports tab or industrial tab, along side Methanol or refineries), LNG will be at the 'kids' table in demands for slower changes in the pipeline quality.

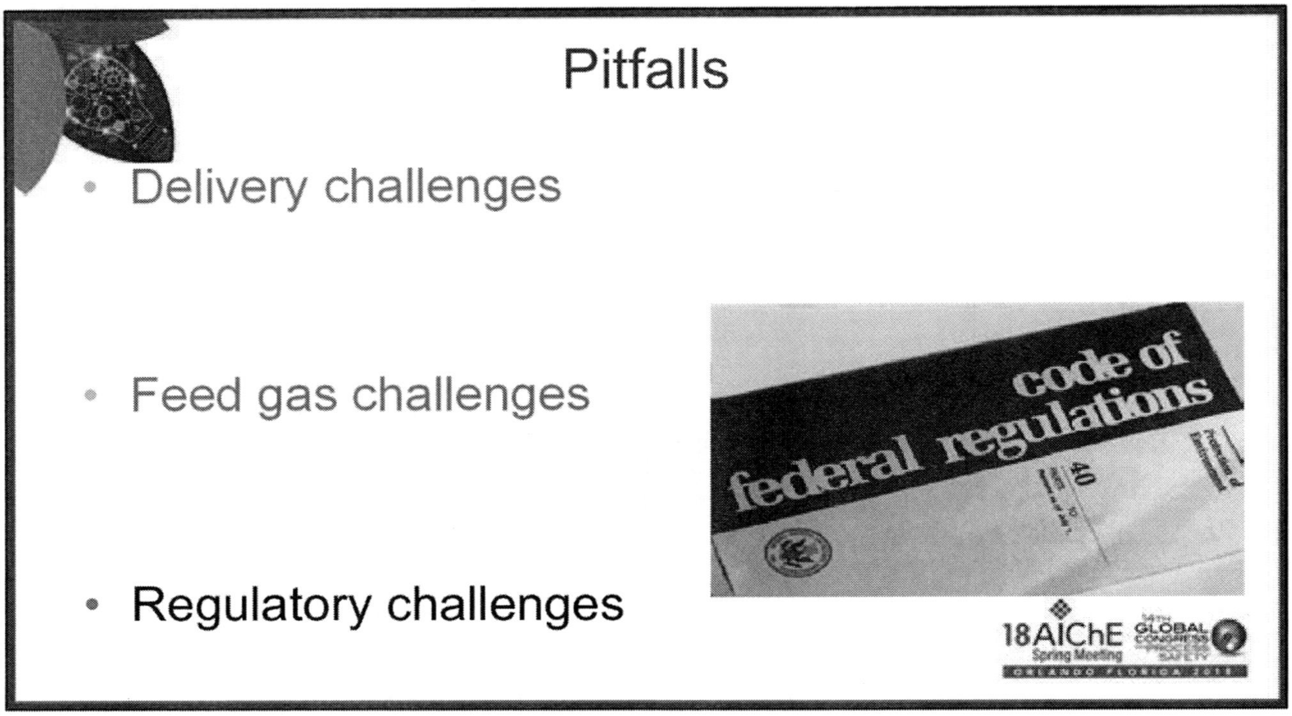

The final pitfall to look at for this presentation is the ubiquitous regulations. Regardless of the country the project will be built, there will be some regulations that will need to be followed. And over the life of the project, some regulations will change; some for the better, some for the worse. How the project is built and executed can determine how later in life regulation changes will affect the plant.

Regulatory Challenges

- Originally built, plant's equipment is 'right sized'
- New requirements may mean updated equipment & designs
  - Ex: Flare monitoring
- Equipment may now be undersized for current production

too big too small

18AIChE Spring Meeting GLOBAL CONGRESS ON PROCESS SAFETY

Regulatory issues can arise when the plant is attempting to increase production. This can also arise due to changes in the regulatory bodies. Both are challenges that can impact plant production. While having items large enough to handle some future expansion, too much can increase CAPEX and sink a project. How do you balance the potential for changes in the future?

Regulatory Challenges

- Current design
  - Right sized, i.e. flare, power, thermal oxidizer
- Future expansion (5, 10, 15 years)
  - Extra capacity?

18AIChE Spring Meeting

Examples of challenges with ethylene producers and refining. Before, flares did not require monitoring for fuel gas heating value. Now they require constant monitoring to adjust fuel gas and steam to ensure smokeless flare operations and a minimum of 200 BTU/scf heating value. With this monitoring comes new control valves, control loops, monitoring stations and testing lines. Do you have space for a monitoring room at the base of the flare (or close by), do you have space to move and build a new flare if you need to? This rule applied to all sites, regardless of new or modified construction. Even if you don't expand (or shrink) you need to consider the potential for regulatory changes and the space requirements.

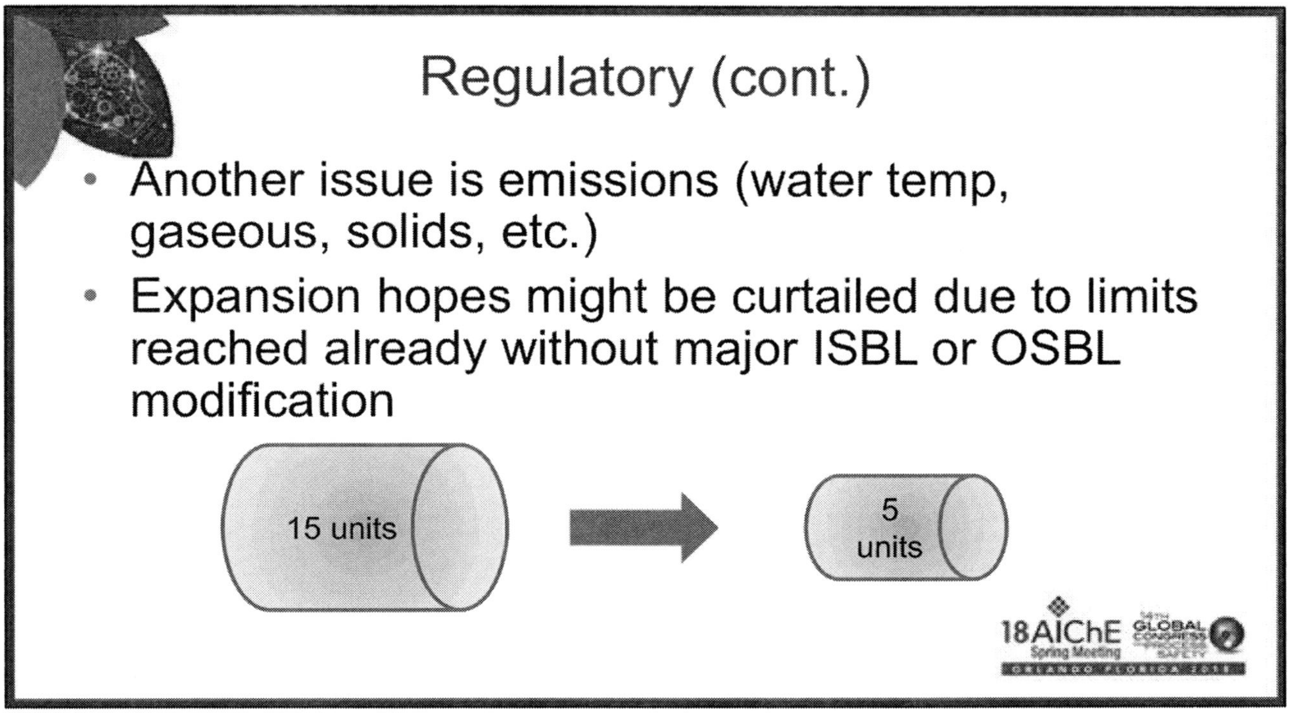

If the project uses river water for cooling, discharge temperatures and salinity may change over the course of the project's life. The plant may have been originally designed for 15 to 10ºF (8 to 5ºC) temperature rise in water discharged. Later, this can be changed to 5ºF (2.5ºC) total rise. Can your plant handle the change? Can you move the water? Does this change your cycles of concentration?

The same issues can be applied to other emissions, such as solid waste, air and liquid discharge.

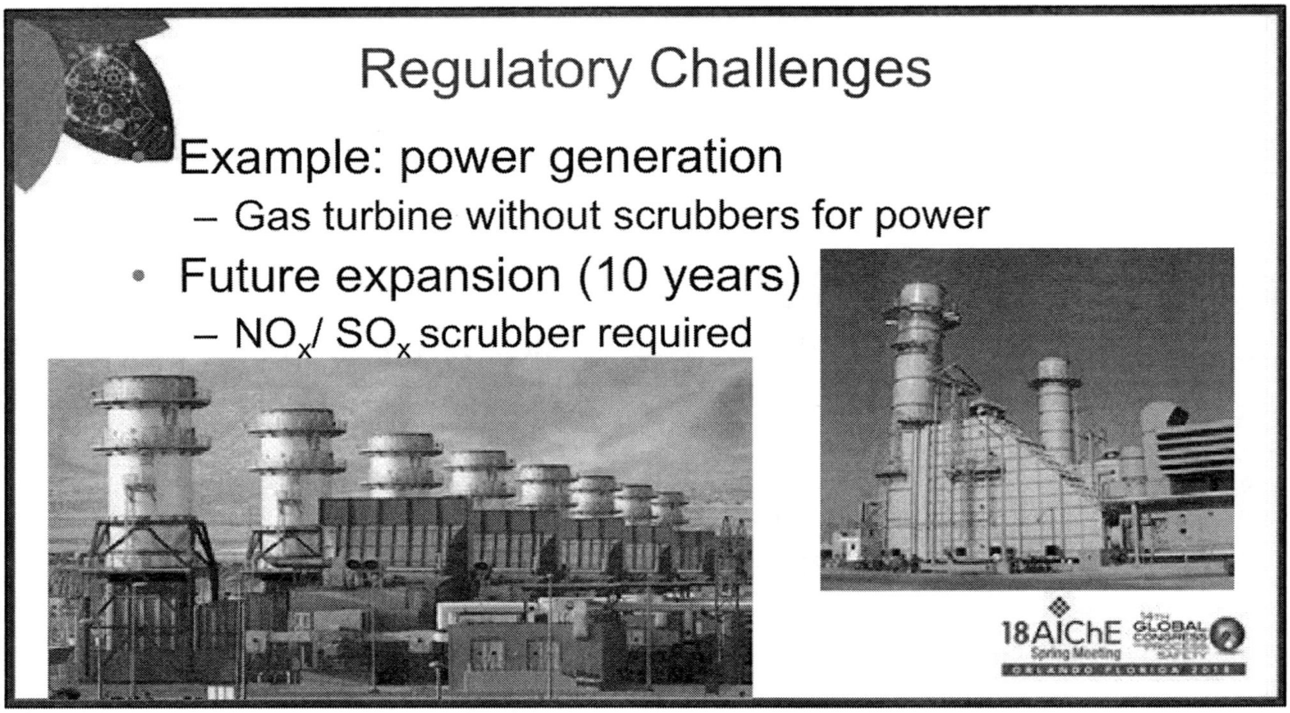

Originally when the plant was built, the power generation on-site burned gas and emitted the exhaust with little to no outlet changes (scrubbing, waste heat recovery, etc.) After several years, new requirements show that the plant needs to reduce air emissions and the plant needs more power for a small addition. A great way to get power would be waste heat recovery, in addition to adding a scrubber. Is there space for the new footprint for the power site? Can you get a new permit? Do you need to rip out the old generator and build new? Can you retrofit? These are issues that identified early on can help the plant and project stay viable over the years.

Added to potential challenges are the air permits, land space and constructability.

Questions?

Jeffery Gaspard, Ph.D., P.E.
KBR Process Engineer
Jeffery.Gaspard@kbr.com

18AIChE Spring Meeting
14th Global Congress on Process Safety
Orlando, Florida 2018

Project Changes in Gas Monetization: Pitfalls or Success - Abstract

Looking at potential Gas Monetization projects can be a daunting task. What can make them more daunting is the prospect of changes in the development and operation of the project. Some changes can be managed during the life of the project: some raw material sources, contract price for finished goods, etc. But other changes can drastically change the project; some move the project from viable to uneconomical. Most of these changes are out of the control of the operator, but with proper planning, some changes can be mitigated during the design phase of a project. This presentation will talk about some considerations that need to be discussed and planned for during the early stages of a project in order to maximize the success of the project.

18AIChE Spring Meeting

STRATEGIES FOR LNG PROJECT DEVELOPMENT

Arif Habibullah, P.E.,
Technical Advisor, Los Angeles, California
Yoshitsugi Kikkawa,
Dr. of Engineering, a Fellow Member of AIChE
Yokohama, Japan

ABSTRACT

As a result of the world's growing supply of LNG, a looming projected glut and weaker energy pricing, marginal projects will be deferred but will continue to be evaluated for improved economics in the future. In addition, due to cost overruns and delays in recent LNG projects, unique conceptual design strategies are needed to assess new projects to improve their economic viability and rate of return.

This paper highlights conceptual design strategies for improving total project economics for all areas from sub-sea, to offshore and onshore development. Strategies are presented to identify areas for reducing costs and enhancing project definition, supported by selective benchmarks derived from recent LNG projects. The goal is to enhance project rate of return and improve total project definition to avoid cost overruns and schedule delays, and help make marginal projects viable.

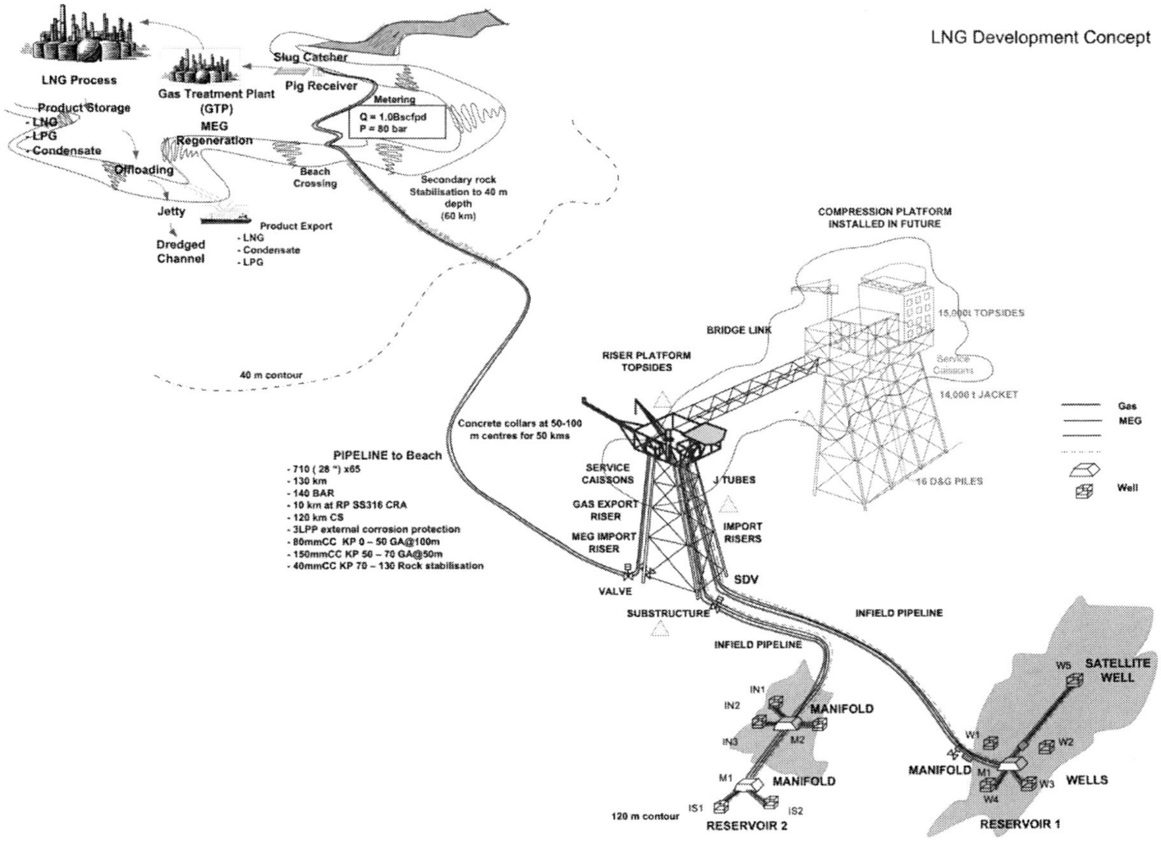

INTRODUCTION

In order to fully evaluate project viability and assess and identify areas for improving the total project economics, a holistic approach is needed that encompasses all components from the sub-sea to offshore and onshore facilities. As such, a clear understanding of each component and its contribution to the total project cost is needed in order to properly identify areas for improvement that will result in enhancing the project economics.

Recent LNG projects, e.g. Gorgon and Wheatstone,have experienced substantial cost overruns and schedule delays; e.g. Gorgon was budgeted at $37B when approved in 2009 but costs escalated to $54B, and first LNG in 2016 was 2 years behind original schedule.Wheatstone cost $34B, $5B more than the originally budgeted $29B and was 6 months behind schedule. Chevron CEO stated the reason for the cost overruns was lack of the needed additional engineering prior to FID.

Our objective is to highlight a few strategies that could improve project economics and provide better project definition in order to avoid cost overruns and schedule delays.

LNG Development Case Study

A schematic of a LNG Development Concept is shown on the cover page forms the basis for the issues discussed in this paper.

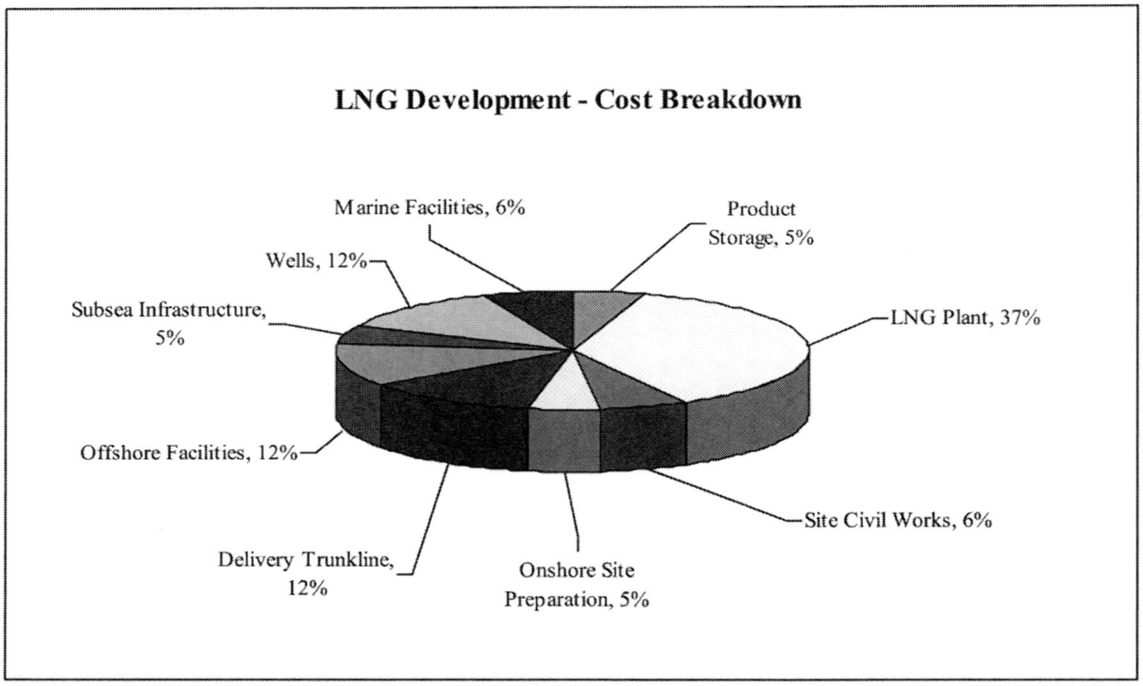

Figure 1 - LNG Development Cost Breakdown

REVIEW OF SITE SELECTION

Site Identification

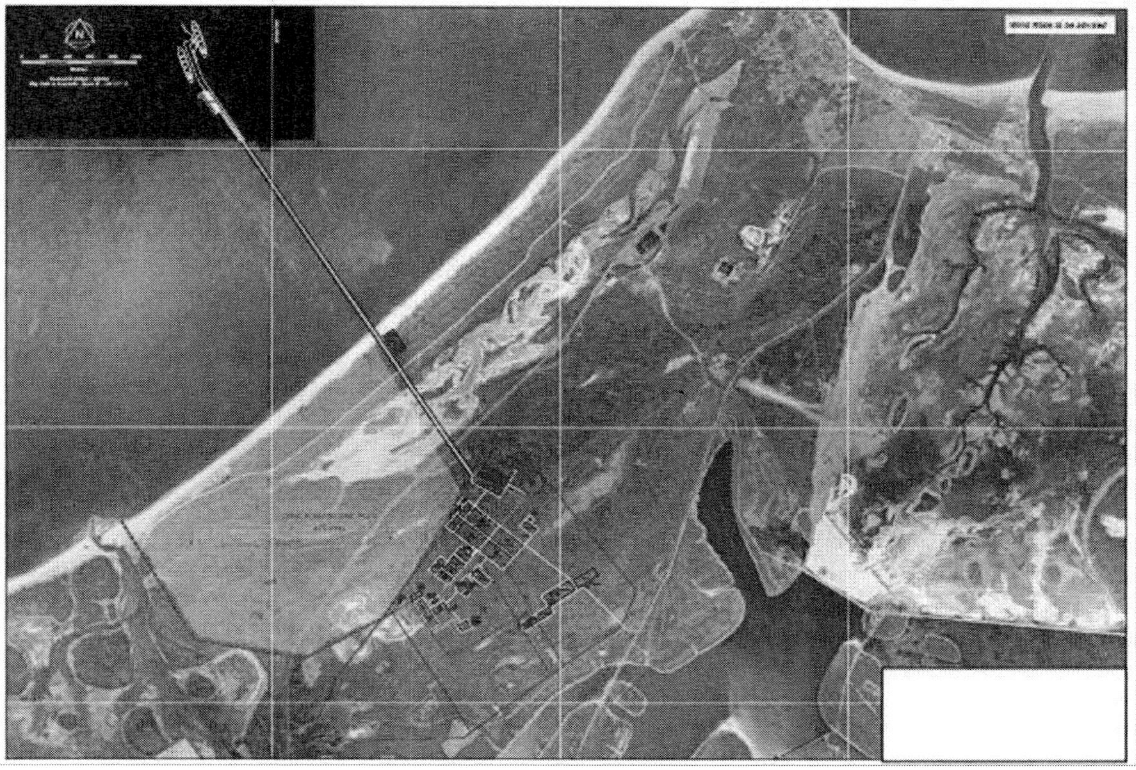

Figure 2 – LNG Site Layout

Assessing a number of selection criteria normally drives the site identification. Once a site has been nominally chosen, a rigorous review utilizing a multi-discipline team may prevent a decision that the owner will regret until the end of operations.
Some of the site identification issues may impact project success include:

- Fully understanding government preferences in order to minimize approval time
- Distance from reservoir to onshore site and impact on technology limits
- Indigenous (Native) title and land ownership
- Environmental Constraints
- Need for a deep water harbor and unrestricted shipping access
- Impact of site works and material availability on project cost
- Access to supporting infrastructure (e.g. Airport, Rail)

Experience has shown that a well-chosen site may reduce site preparation and marine costs (i.e. earthworks and dredging) by up to 5% of total capital cost.

ENVIRONMENTAL ISSUES

Typically there is a resistance to pre-invest prior to project sanction. However, experience would suggest that the availability of accurate environmental data can significantly reduce design risk especially in areas such as subsea, pipelines, offshore structures, onshore foundations and marine facilities.

Environmentally sensitive aspects may include:

- Site for the offshore facilities
- Site for onshore plant
- Trunkline route
- Trunkline beach crossing
- LNG tanker operations
- Dredging
- LNG jetty
- Construction rock availability and quality

Poor definition of environmental criteria can result in extensive rework in areas such as offshore foundations, trunkline routing and operations (i.e. quarantine management).

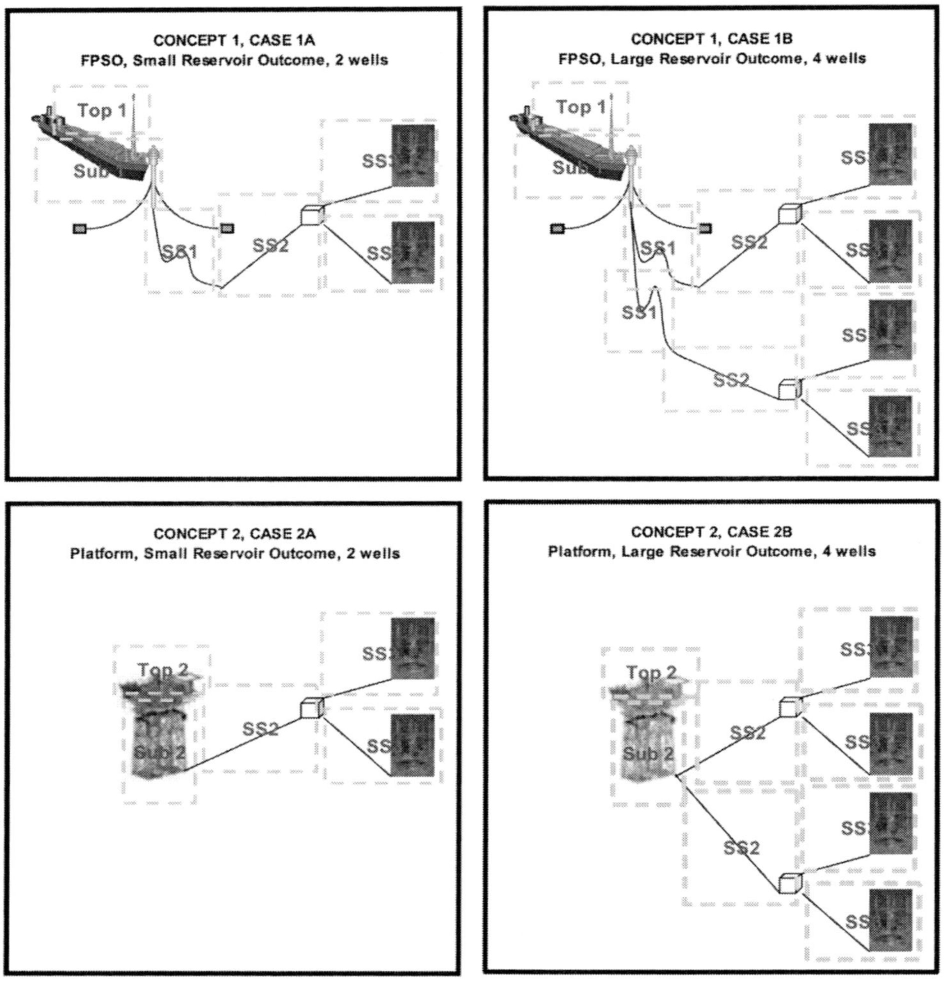

Figure 3- Key Components and Using Building Blocks

OFFSHORE DEVELOPMENT

Drilling

Well costs can make up 12-15% (see Figure 1) of the complete development depending on geographic region, water depth, reservoir characteristics, number required and installation technique. Selecting the optimal installation technique can saveup to 50% of the well installation costs. Typically, the owner must make an early decision on whether to usea full subsea development versus having an offshore production facility.
If an offshore facility is considered, then it is necessary to evaluate:

- Drilling "wet wells" using a Semi Submersible (Semi-Sub) Drilling Rig
- Drilling "dry wells" by Tender Assist (TAD)
- Drilling"dry wells" from facility based integrated drilling module
- Drilling a combination of the previous three items

Note: *Wet wells are subsea wells, which utilize subsea Christmas trees while dry wells are installed on the platform topsides.*

A well-developed component based (building blocks) approach (See Figure 3), has been shown to enable quick response to changing reservoir and well constraints and quickly allow reconfiguration of options.

Subsea Recovery Systems

A full subsea recovery system removes the need for expensive offshore real estate and the ongoing associated operating cost. This needs to be balanced against reduced access to wells and reduced flexibility in process design. The decision is often closely linked to the results from flow assurance. Options include:

- Complete subsea development with remote controls at onshore facility
- Complete subsea development with spar supported controls
- Subsea infrastructure complimented by offshore facility (See Figure 4)
- A variety of subsea configurations such as ring main and daisy chain

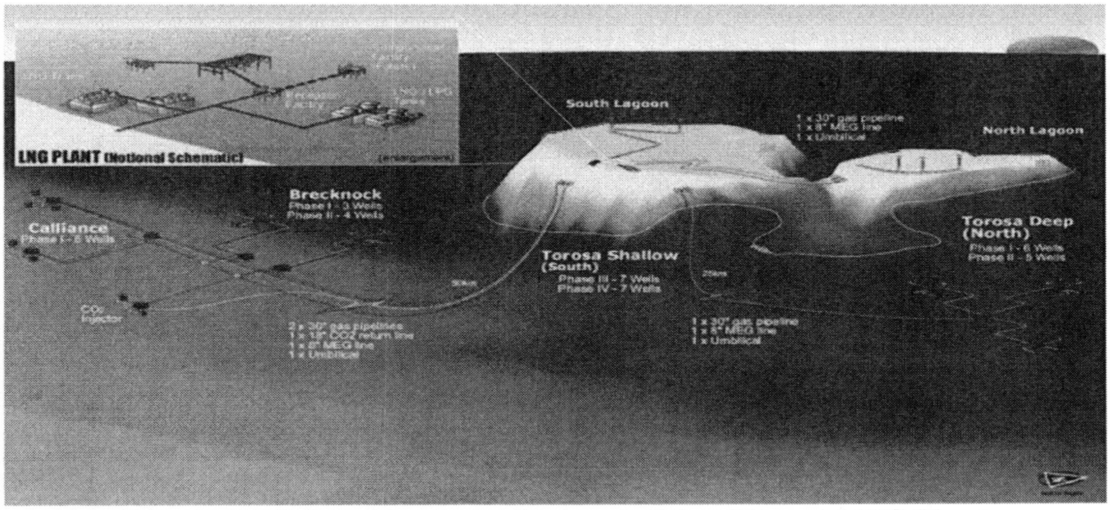

Figure 4 – Subsea Recovery System to Island Based LNG Plant

Offshore Facilities

Substructure

Offshore facilities can be either floating units or structures fixed to the seabed.Several Floating LNG (FLNG) concepts have been in development to-date. FLNG places the LNG plant facilities at the production site removing the need for an expensive trunkline, onshore receiving facilities, onshore process facilities, onshore storage and export facilities.

An FPSO can be used in lieu of a platform and is attractive if the field has a large amount of condensate as the condensate can be recovered, stored and exported directly from the field. This leaves dry gas to be exported to the onshore LNG plant.

Other floating options are used in deeper water developments where conventional "bottom fixed" structures would be too expensive or technically infeasible.

There has been a drive over the last 10-15 years to fabricate and construct facilities in low cost centers such as South Korea or China to reduce costs.

The following Floating and Bottom Fixed options provide a sample of what is available;

- Floating
    - Floating LNG Facility (FLNG)
    - Floating, Production, Storage & Offloading Facility (FPSO)
    - Semi-Submersible (SemiSub)
    - Tension Leg Platform (TLP)

- Bottom Fixed
    - Steel Jacket
    - Jack-up
    - Concrete
    - Hybrid
    - Proprietary (custom made and installed – i.e. "float outs")

Topside Facilities

There are many potential decisions with topsides layout and phasing which can impact the overall CAPEX of offshore facilities. One of the primary decisions is to how to handle future compression. As the pressure of a gas field declines, compression is required to maintain feed gas flow. The decision on the compression installation and phasing needs to address:

- Cost to accommodate future gas compression on the start-up facility vs construction of a future additional facility (pre-investment)
- How this decision affects the configuration of the subsea recovery system
- Impact on operations

Decision to pre-invest is largely dependent on the pressure decline period. Typically if compression is required within 6-8 years after start-up then pre-investment may be an attractive option.

Trunkline

The function of the trunkline is to transport gas from the offshore field to the LNG Plant. The trunkline can be operated wet, dry or wet with MEG injection. Operating wet reduces offshore equipment and simplifies offshore facilities making it possible to operate an "unmanned" facility. The downside is higher pipeline material costs, cost of chemical injection, regular pigging and the potential for flow assurance issues such as hydrate formation. As installed trunklines can cost as much as $US10 million/mile, then small improvements can result in large savings.

Dry pipelines typically use carbon steel but require costly separation, dehydration and dew point control facilities offshore. MEG injection reduces some of the flow assurance issues but requires a costly recovery and recirculation plant either offshore or onshore.

For field life of over 15 years, a dry trunkline concept will most likely provide a lower lifecycle cost than a wet trunkline option. A primary input into this decision is the impact lost production in the LNG facility.

In certain parts of the world, the process discussion can be complicated by the necessity to provide extensive stabilization for the pipeline. While providing protection from the environment, the stabilization media can significantly alter the pipeline temperature profile and therefore, the physical properties of the production fluids.

ONSHORE DEVELOPMENT

The Onshore facility represents a significant part of the total project investment (~60%) and therefore, offers much opportunity for cost reduction and improvement in project life cycle economics[5].Refer to Figure 1.

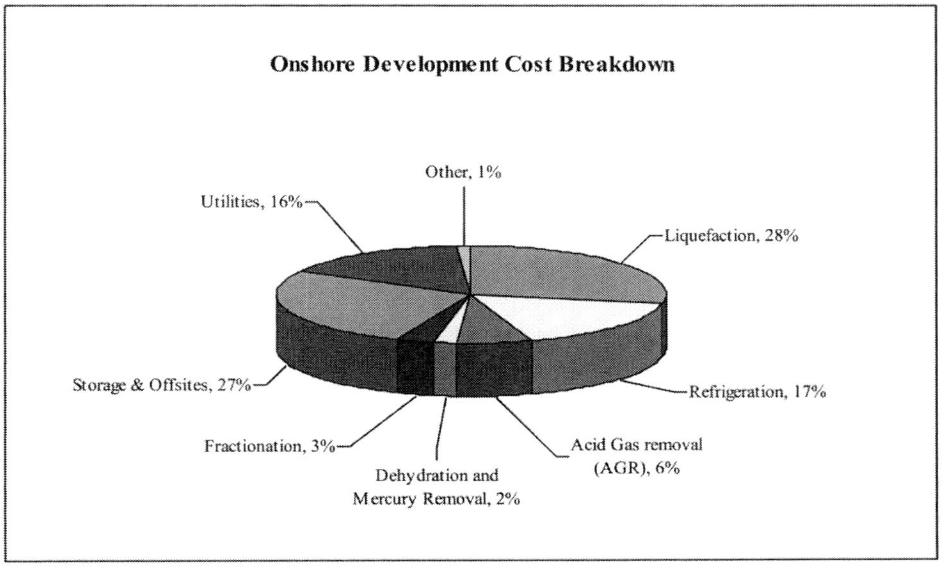

Figure 5 - Onshore Development cost breakdown

LNG train size and number of trains

Selecting a train size and number of trains is a complex issue and choices will vary widely depending on various project specific considerations, including available gas supply and future sales demand.

Over the past decades a trend towards increasing train size is clearly established driven by cost savings via economies of scale. The historical growth in train size is well documented and has been frequently published in various technical journals. Multiple large trains of 7.8Mtpa capacity were installed in Qatar [3,4]. If this trend continues in the future or not is debatable, but it's likely that future LNG train designs will fall within three groups with nominal capacities of ≤3, ≤5 and ≤8Mta. The first group will cater to limited gas supply or sales environment, the next to higher gas supply and sales and the mega trains to large complexes supported by large reservoir and serving distant global markets. So far it seems unlikely if any mega trains will be installed, judging from current LNG project slate. One exception is Qatar where four additional mega trains are planned as part of their expansion plan to increase capacity from 77 to 100Mta.

Reservoir size must support the plant capacity through the duration of the project life. For a given plant capacity the size and number of trains depends on many factors including some of those listed below:

- Reservoir size
- Site conditions including plot space
- Existing infrastructure
- Phased-in capacity to meet LNG market and future expected demand (initially build smaller trains and add trains as demand grows)
- Greenfield or expansion of existing facility, e.g. replicating existing trains
- Economies of scale (e.g. a single train facility saves over 10% of capital cost of 2x50% trains)
- Provisions for future expansion in a base case design
- Major Equipment size limitations (e.g. AGR absorber, main cryogenic exchanger, C3 compressor)
- Vendor base to build large equipment
- Transportation constraints
- In a modularized plant the maximum transportable module size can limit train size, depending on site conditions

Modularization

The costs of modular units are generally 10-15% higher than field erected units since modules are designed to withstand the rigors of shipping and require more structural steel. However, performing the work at the fabrication site instead of the field often offsets higher cost of modularization. It reduces the overall project cost due to lower field construction costs and shorter construction schedule[1].

Figure 6 - Woodside Train V Expansion Project module (35-1800tons module range) 18,000 tons total module weight.

Few LNG modularization concepts have been applied to-date. However, alternate modularization approaches will continue to be developed in order to further reduce costs.

The Snovhit project (See Figure 7) was based on stick building the LNG train on a custom designed barge in Spain and then shipped to site on a heavy lift vessel. This strategy reduced the construction schedule particularly due to the site's narrow construction window. An alternative approach may involve building the barge and modules *concurrently* prior to final module assembly, resulting in further schedule reduction.

The Woodside Train V expansion project (See Figure 6) in Australia took an existing plant layout and modularized it at a fabrication site in Batam, Indonesia, for savings in site construction labor costs. An alternative approach may involve developing a layout that is customized for modular design to further reduce costs.

The Gorgon Project in Australia is one of the only LNG projects in the world for which modularized construction was designed from the outset. However, the decision to adopt a modular construction strategy was primarily taken to reduce environmental impact to Barrow Islandduring the construction phase.

Along with the above unique and one of a kind design strategy, a comprehensive Quarantine management plan was implemented to prevent the introduction of non-indigenous animals and plants that could potentially establish on the island, predate native species or compete with them for food, water and shelter.

Modularization can reduce costs, however, site-specificissuesshould be addressed:

- Environmental regulation- including cost of environmental issues related to bringing in modules to construction site
- Where lengthy permitting process is involved the project schedule can be reduced by early start of work at the fabrication site
- Local skilled labor availability
- Site access
- Plot space limitation
- Potential of weather related delays and issues with construction window
- Extent of dredging requirements to bring large modules in

Larger modules can reduce costs.

Benefits of large vs small modules:

- Fewer process systems are split between modules
- Fewer inter-module connections
- Man hours/ton are less for fabrication and installation
- Less engineering
- Less plot area is used
- Fewer MCC/instrument rooms
- Hydrotesting and PCO work is easier and more complete
- Fewer foundations
- More hours spent at fab site; therefore better quality
- Larger or taller equipment is more easily accommodated
- Reduction in schedule at installation site (faster start-up)

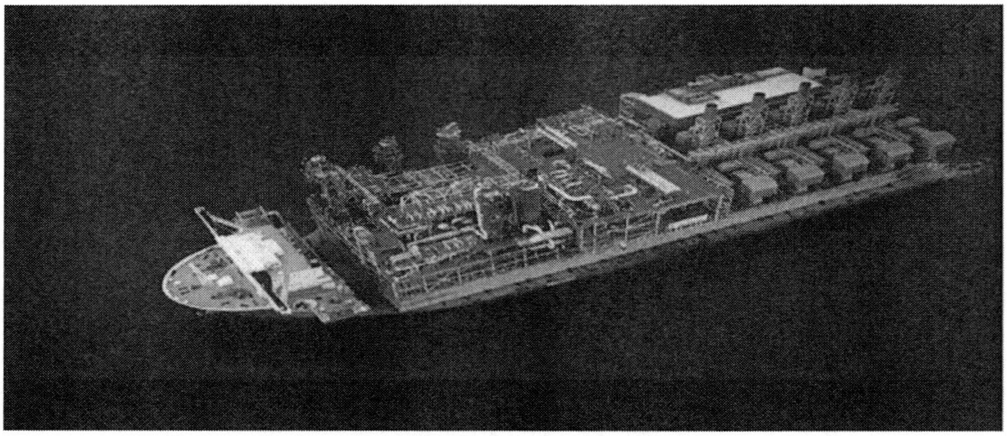

Figure 7 - Snohvit LNG barge (35,000 ton module)

Benefits of Super Modules: (2000-5600 tons)

- Substantial reduction in site labor hours
- Safer- less site exposure
- Wider contractor pool for remaining work scope

- Up to 50% reduction in plot space
- Significant CAPEX savings
- Site directs substantially reduced
- Reduced camp requirements

As summarized in the referenced study [1] titled "Modular vs Stick Built LNG Plant" the following conclusions were highlighted, specific to that particular project:

1. The total direct cost for a modular LNG plant was 10.5% less than stick built.
2. The TIC for modular case was estimated to be 13.5% less than stick built.
3. Estimated contingency for modular case was15% less than stick built. This is a reflection of the reduced risk and uncertainty on project delivery.
4. The overall weight for a modular plant was 20% greater than for stick built due to the increased steel required for modular construction.
5. Modular design requires higher quality of early engineering and a larger management effort (4% higher).
6. Fabrication hours increase by about 29% but overall fabrication costs are 28% lower due to lower fabrication yard labor rate.
7. A reduction in construction man- hours of up to 15% due to significant reduction in field construction scope.
8. Construction camp costs were41% lower than stick built.

Other reasons, besides costs, for considering a modular approach are:

- Lowest exposure and cost/schedule impact if longer periods of adverse weatherconditions are experienced.
- Reduced pressure on social infrastructure at job site.
- Highest potential for cost optimization by maximizing areas of modularization.
- Utilizes broader labor market and not fully dependent on local labor conditions.
- Enhances competitive bidding especially for module fabrication.
- Reduces infrastructure requirements at job site.
- Reduces job site manpower demands and related construction facilities.

Plot Space Reduction

A smaller plant footprint can reduce costs. Reduction in footprint may also be necessary when adding LNG trains to an existing complex within limited space. Either way, various strategies include:

- Common facilities- utilities are often located in a common area. However, it may be possible to combine other units to service multiple trains, e.g. inlet facilities, common fractionation plant, LPG treating, etc.
- Stacking of equipment - e.g. refrigeration suction scrubbers, propane chillers (including multi-level structure)
- Use of proprietary internals (e.g. centrifugal devices) to reduce vessel diameters

- Use of new low finned and double enhanced tube technology (using nucleate boiling structures for small delta Ts), for significant reduction in C3/MCRchiller and condenser diameters and length
- Reduce controlling flare load and hence radiation circle by eliminating C3 compressor blocked discharge case, (see Flare Minimization)
- Combine all flare stacks in a derrick retractable flare system, including spare stack
- Use inlet gas heating for hydrate control vs e.g. TEG
- Consider DLN (Dry Low NOx) vs SCR which requires significant plot space and higher OPEX and CAPEX. DLN also improves plant availability
- Use GT exhaust heat for regenerating molesieves (eliminates regeneration fired heaters); use newer technology utilizing combo molesieve beds for mercury removal

Motor Driven LNG plant

A motor driven LNG plant increases thermal efficiency of the facility providing higher LNG production. The larger and more efficient GTs in a centralized power plant (possibly CCGT) can provide an overall efficiency increase of 5-10%. This could translate into an incremental LNG production up to 0.5%, depending on project specific requirements and conditions.

Large motor drivers are now being considered as an option for driving refrigeration compression. Statoil's Snohvit LNG plant opted for 65MW motors for refrigeration compressors powered by a centralized power plant consisting of LM6000 GTGs. However, the facility is connected to the grid for back up which is unusual for most LNG projects under consideration. The mega LNG trains in Qatar use Frame 9Es with 45MW(rated) motor/generators for refrigeration compression(3,4).

The commercial benefits of a motor driven plant are:

- Higher production efficiency (i.e. more LNG production)
- Higher thermal efficiency
- Lower CO2 emissions
- Shorter delivery schedule of motors vs GTs reduces overall schedule
- Motors are better suited for a modular designs, hence lower installed cost

CAPEX for the electric motor driven plant option is higher, primarily due to the addition of a large power plant. However, there is a net increase in NPV due to the higher availability compared to a GT driven plant, and a shorter project schedule. These two factors alone often outweigh the higher cost of the larger power plant.

Improving Plant Efficiency via Aeroderivative GTs

The two areas that influence the LNG plant efficiency are turbo-machinery and cryogenic heat exchange. Given the maturity of LNG liquefaction processes, there is little room for further tightening ofexchanger temperature approaches(2,6).

That leaves two areas that significantly impact plant efficiency:
1. Refrigeration compressors
2. GT drivers.

Compressor efficiencies are already in the high 80s, therefore, selection of the GT will determine the thermal efficiency and CO2 emissions for the liquefaction turbo-machinery. With increasing emphasis on reducing greenhouse gas and the higher value of fuel, the more efficient aeroderivative GTs offer commercial benefits. With a thermal efficiency of over 40% comes a reduction in CO2 emissions. CO2 emissions are one-third less per HP generated with an aeroderivative vs a Frame unit (see Table-1).

Table 1 - Compressor Driver Options

Type	ISO Power MW	Efficiency %	Relative CO2 Emissions	Speed RPM	Weight kg
Frame 5	32.5	29.4	1.0	4,670	110,000
Frame 7	87.3	33.1	0.93	3,600	121,000
Frame 9	130.0	34.6	0.87	3,000	217,500
Electric Motors*	65.0	50.9	0.58	Variable	-
Aeroderivatives	43.9	41.9	0.71	3,600	31,000

* CC+ VSD Higher Availability

Pros

- If the gas is pipeline limited or otherwise constrained there is a clear benefit for higher efficiency GTs (any fuel savings add to LNG production)
- Aeroderivatives are a third the weight and a quarter the foot print of an industrial GT
- Aeroderivative GTs achieve higher thermal efficiencies than industrial GTs
- Higher efficiency of aeroderivatives corresponds to >3% increase in overall plant efficiency
- Plant availability is higher since it's possible to change out a complete GT within 48hrs vs 15days or more needed for industrial GTs
- Reduction in fuel consumption reduces CO2 emissions.
- Published data for CO2 emissions for various LNG projects have been compared and values in the 0.25-0.37 tpa CO2/ton LNG produced have been reported [5,6].
- Using a more realistic fuel value of say $3-$5/mmscf (vs the typically applied $0.50/mmscf 'stranded gas' fuel value) will substantially improve life cycle costs for options with higher efficiency

Cons:

- higher CAPEX per BHP.
- No large power frame such as Frame 7, 9 is available

Issues:

- Even the largest aeroderivative GT proven in mechanical drive service (LM6000) has about ½ the power of a Fr7
- Aerodervatives require more maintenance (time between major overhaul is 24000hrsvs 48000hrs for industrials) and require higher-pressure fuel gas.

RAM Analysis

A conceptual RAM analysis can predict the facility production performance and identify key contributors to production loss, by unit. This type of preliminary analysis helps identify areas with the highest potential for increasing Production Efficiency (PE) and hence more LNG production. An overall analysis encompassing offshore and onshore facilities is typically carried out; for this discussion we are addressing the onshore facility only.

Results from an analysis of this type varies widely and depends on the technology applied, plant configuration, sparing philosophy and type of compression drivers, etc. For GT driven LNG plants studies show that the refrigeration drivers are the largest contributors of the losses through scheduled and unscheduled outages. Up to about 2/3rd of the total losses can be attributed to refrigeration drivers.

Therefore, any reduction in the durations needed for scheduled turbine inspections can improve PE substantially. Using the owner's plant inspection history vs manufacturer's recommendations should be reviewed for a more site specific and realistic assessment and a possible reduction in scheduled downtime.

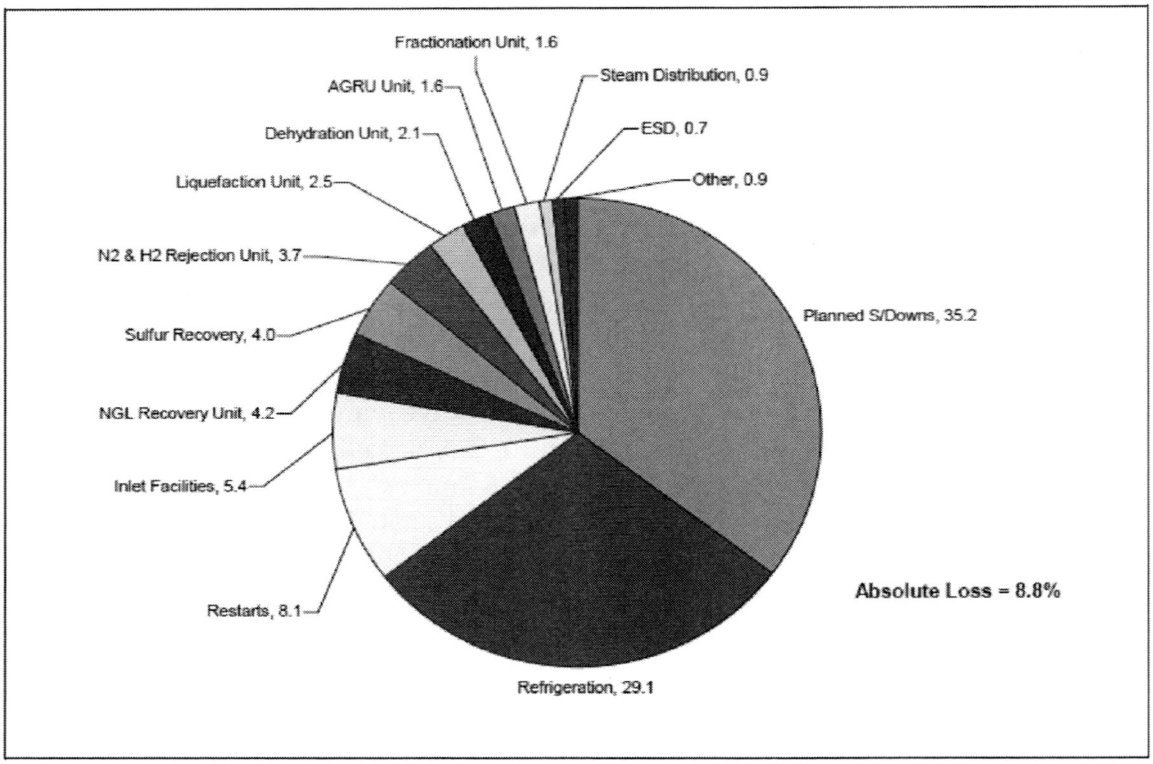

Key Contributors to Production Loss (%)

In summary, some areas for increasing availability are:

- For GT driven plants the production efficiency is driven by the GT turbine drivers
- The largest production loss is caused by the scheduled GT turbine maintenance schedule which governs the performance of the LNG plant
- The second largest contributors to the losses are the unscheduled failures of the GTs and compressors (up to 70% of losses due to GT, 25% due to compressors) that drive the refrigeration trains
- The third largest contributors are the restart losses after a plant trip. Restart duration can be reduced by using software such as AP-AutoCool™ linked with DCS.
- Annual production efficiency varies year by year and is highest during the years with no gas turbine maintenance and can drop by up to 5% or more during years with Major Inspection(MI), and up to 3% during the Combustion Inspections (CI) and Hot Gas Path Inspections (HGPI).
- Reduction in scheduled maintenance durationsfor both HGPI and MI by say 5 days, PE can potentially be increased by ~0.4%
- Consider use of GEExtendor™ kits, if applicable, to increase CI intervals to twice as long; e.g. PE increase of up to 0.8%
- Single train restart time is dominated by the required cool down time and a reduction can increase PE by up to 0.2-0.3%
- 'What-if' scenarios should be run to evaluate impact of sparing selective equipment on PE. For example, sparing the off-gas compressor in the inlet facilitycan result in PE increase of up to 0.5%.

Further increase in plant availability is possible:

- Review if flaring is acceptable during FG compressor trip vs a train trip
- Review if SRU trip and acid gas flaring is acceptable vs a train trip

Flare Minimization

With increased emphasis on taxing CO2 emissions and the high cost of fuel, flare minimization reduces project life cycle cost. Flaring reduction can potentially be achieved by the following considerations:

- Review of various HIPS applications have been successfully employed in the LNG industry for reducing costs without jeopardizing overall safety [7]
- Design the facility for 2:1 or higher turndown to minimize flaring
- Incorporate sufficiently high % of fuel-from-feed in the fuel balance to allow for better control and trimming of fuel to maintainplant fuel balance without excess flaring
- Design the fuel gas compressor with a good margin to allow for future increase in end flash gas production (as throughput increases) to avoid flaring excess fuel gas

- Provide means to recycle flared gas to front end of the plant (e.g. collect derime gas and recycle to dehydration unit inlet via fuel gas compressor); also recycle gas during pre-cooldown instead of flaring
- Consider increasing design pressure of C3 refrigeration compressor casing (often the controlling case for flare sizing) to reduce flare size (by up to half) during blocked discharge scenario
- Consider routing streams normally going to flare, where possible, to fuel systems
- Upgrade specs to include double isolation on all connections to flare to minimize losses due to leaks; include 150# systems which would normally have single isolation
- Critical control valves fitted with isolation to allow for removal of malfunctioning valve without purging entire unit
- If sufficient driver HP is available, consider the possibility of a full pressure restart/or partial depressurization for faster restart of the liquefaction plant after a trip; also increases plant availability

Nitrogen Rejection

Typical nitrogen spec of 1mol% is often met using an end flash, however, any further reduction in nitrogen content allows shipment of more LNG, say 2 or 3 additional cargoes per year. In addition, today's large ships are equipped with re-liquefaction plants for which the higher nitrogen content is not optimum.

In summary, the modest incremental capital investment in a stripper column to further reduce nitrogen content and increases revenue and:

- High Btu, low N2 LNG provides greater annual revenue
- Increases LNG shipping efficiency and produces more LNG
- Provides incremental condensate sales revenue
- Newer ships (e.g. Q-flex LNG carriers) are fitted with re-liquefaction units, therefore more cargo is delivered with each voyage
- Provides better control on the amount of N2 in fuel gas (important for meeting gas quality for DLN combustors)

Issues:

- No incremental LNG if feed gas limited
- Fuel gas from the column has a higher nitrogen content and may require blending with other sources to meet DLN requirements

TAKEAWAYS

As a result of the world's growing supply of LNG, the projected looming LNG glut and continuing low energy pricing, marginal projects will be deferred but will continue to be evaluated for improved economics in the future. In addition, due to cost overruns and delays in some recent LNG projects, unique conceptual design strategies are needed to re-assess new projects to improve their economic viability and rate of return.

Developing strategies for improving project economics requires revisiting project design concepts of ***ALL*** components including:

1. Sub-sea,
2. Offshore and,
3. Onshore development.

A total project assessment capability is required to properly identify areas of improvement with potential for enhancing project economics.

To improve total project economics and avoid cost overruns and schedule delays requires a thorough assessment of the design criteria and its impacts, as summarized below:

Subsea &Offshore Development

- **Project Site Selection** impacts dredging costs, site prep, earthworks/infrastructure (5% of TIC)
- **Accurate Environmental Data** reduces design risks, e.g. better defines site for offshore facilities, trunkline route, dredging, LNG Jetty
- **Well Installation Costs** (12-15% of TIC) can be reduced by evaluating use of full subsea development vs offshore production facility
- **A Full Subsea Recovery System** removes the need for costly offshore real estate and ongoing operating costs; this should be balanced against reduced access to wells and reduced flexibility
- **Substructure –FLNG Concept** places the LNG plant at the production site and eliminates the need for a trunkline, onshore receiving, storage and export facilities
- **FPSO** in lieu of a platform is attractive if the field has large amount of condensate, which can then be recovered and exported directly from the field, leaving dry gas to be exported onshore
- **Phase-in Future Compression** as the field declines vs pre-investment in additional facilities
- Evaluate wet trunkline vs dry or wet with MEG injection

Onshore Development

- **Train Size & Number** should be balanced against economies of scale; pre-invest in infrastructure
- **Modularization vs Stick- Built** or a hybrid design, could potentially save 10-15% of TIC
- **Plot Space Reduction** by stacking chillers, use of proprietary vessel internals, eliminate C3 compressor blocked discharge case, combine all flare stacks in a derrick structure
- **Motor Driven Plant** - centralized power plant, could increase LNG production up to 0.5%
- **Aeroderivative GTs** can improve plant efficiency by>3%, reduce carbon footprint by up to 30%
- **Conceptual RAM Analysis** identifies options for increasing LNG production; restarts are one of the largest contributors to production losses, restart at full pressure or with partial depressurization
- **Flare Minimization** reduces project life cycle cost and carbon footprint; design plant for high turndown; consider HIPS; design for high FFF in fuel balance to avoid excess flaring
- **Nitrogen Rejection**, maximize nitrogen stripping to increase LNG shipping efficiency and increases revenue; could provide incremental condensate sales volume.

References

1. Habibullah, A., et al, Modular vs Stick Built LNG Plant, GPA Conference, 2010, San Antonio, Texas.
2. Habibullah, A., et al, Carbon Footprint Reduction Strategies, Gastech2009, Abu Dhabi, UAE.
3. Salisbury, et al, Design, Manufacture and Test Campaigns of the World's largest LNG Refrigeration Compressor Strings, 2007 LNG15 Conference, Barcelona, Spain.
4. Pillarella, et al., LARGE LNG TRAINS: Developing the Optimum Process Cycle
5. C.Yost, et al, Benchmarking study compares LNG costs, OGJ April14, 2003.
6. Jeff, Sawchuk et al, "BP's Big Green Train:Benchmarking Next Generation LNG Plant Designs, Gastech 2005.
7. Thomas, C., et al, High Integrity Protective Systems, OGJ May 8, 2000.

512780 Optimal Synthesis of Integrated Reactive Distillation Systems for Fischer-Tropsch Process

Abstract:

Fischer-Tropsch synthesis (FTS) is an area that is receiving revived interest as a technology alternative to produce transportation fuels as well as chemicals from syngas.

Reactive distillation (RD) combines separation and reaction in a single unit, which is ideal for process intensification systems. As heat removal is an essential task for FT reactors, this paper aims to prove that RD can be an advantageous candidate in driving separation of products from the exothermic reaction process, as well as improving selectivity and enhance product yields.

A steady-state adiabatic reactive distillation model is proposed for the FTS. The calculation of the vapor-liquid equilibrium (VLE) through cubic equation of state is used to describe the phase behavior. Rate expressions for the FT and the water gas shift reactions are taken from literature and expressed in terms of fugacities. Product selectivity of catalysts are implemented based on Anderson-Schulz-Flory (ASF) distribution and experimental data. The mass, equilibrium, heat and summation (MESH) model, which is extended by considering bypass streams for non-reactive trays, is used to integrate column structure blocks. A step-by-step initialization procedure is proposed to accommodate for the complexity of the RD column and the high nonlinearity of the model.

Studies of operating variables are conducted to enlighten optimization over performance of the RD model. A typical low-temperature Fischer-Tropsch (LTFT) process which favors diesel production is implemented in the RD model. Results obtained using the RD for FTS are compared against conventional CSTR models, the results showed that RD would have a potential edge in industrial processes.

Keywords: *Fischer-Tropsch Synthesis, Reactive Distillation, Vaporâ€"Liquid Equilibrium, Kinetic Modeling, Initialization Procedures.*

A Mini-pilot Scale Demonstration of Single-stage Syngas-to-DME Process

Lujie Ye and Sunggyu Lee

Sustainable Energy and Advanced Materials (SEAM) Laboratory

Ohio University

Athens, OH

Extended Abstract

A fully continuous, proof-of-concept mini-pilot scale system for a process engineering study of single-stage syngas-to-dimethylether (DME) process [1] has been designed and fabricated on a self-enclosed mobile platform. In this mini-pilot system, hydrogen, carbon monoxide, carbon dioxide and methane are fed at one side of the pilot; methanol and DME, as products, are collected and analyzed at the other side of the pilot. The mini-pilot system is composed of four seamlessly-integrated principal sections of (1) a feed syngas blending and compression system, (2) a mechanically agitated slurry reactor (MASR) system, (3) a product separation, collection and analysis system, and (4) a data acquisition and control system (**Figure 1**). The entire system can be operated from the outside of the enclosed unit and is in principle remote-controllable.

1) Feed syngas blending and compression system

A feed syngas is blended using separate sources of hydrogen, carbon monoxide, carbon dioxide and methane in this blending system. The flow rates of each gas is metered and controlled by using a mass flow controller (MFC) powered and controlled by Brooks four channel box (model 0254AB1B11A). The composition of the blended syngas is adjusted proportionally via using hydrogen as the master flow. The obtained syngas with stable composition is then pressurized by a Haskel pneumatic booster pump and stored in a gas storage tank, which is installed with a rupture disc designed to actuate at 4000psi at 22°C. The mixed syngas is sent to the reactor and/or to the gas chromatograph (GC) to analyze the volume percentage of each component. For the operational safety of the system, combustible gas and carbon monoxide monitors with audible and visible alarms are mounted next to the gas cylinders.

2) Mechanically agitated slurry reactor (MASR) system

The heart of the reactor system is a one-liter Parr Systems® bolt-closure stainless steel reactor vessel with an air-driven magnetic drive assembly installed (Model 4520 Bench Top Reactor). A 4848 reactor controller is equipped with the reactor to control the temperature with built-in Type-J thermocouples. The maximum allowable working pressure and temperature for this reactor are 1900psi and 350°C. A rupture disk is installed to vent at 2000psi at 22°C. The flow rate of the blended syngas is adjusted by another Brooks® mass flow controller (equipped with Brooks® Model 0251). After passing through a forward pressure regulator (FPR), the syngas is introduced to the reactor from the top. In consideration of safety, a nitrogen tank is connected with a power to close solenoid valve. In the case of power failure, therefore, the system will be automatically flushed with nitrogen. Two combustible gas monitors and one carbon monoxide monitor are installed near the reactor to detect and warn any potential leakage.

3) Product separation, collection and analysis system

After the reaction, the product and unreacted gaseous species flow out of the reactor through a product line, which leads to a condenser, i.e., a phase separator. After cooling down to a temperature substantially lower than 64.7C (boiling temperature of methanol), vapor phase methanol and water will condense to the liquid phase, which can be separated and collected. The liquid phase product is sent to GC, which is equipped with a flame ionization detector (FID), to analyze the content of the methanol. The gaseous phase comprising of DME and unreacted gases is sent to GC-thermal conductivity detector (TCD) for analysis.

4) Data acquisition and control system

The developed LabVIEW® data acquisition and control system consists of three platforms: 1) The temperatures and pressures monitor and control platform, 2) remote solenoid valves control platform, and 3) safety monitor and alarm platform, which includes five gas monitoring alarms, over-temperature alarm and over-pressure alarm. This digital system allows a real-time and remote data acquisition and control for the entire mini-pilot system.

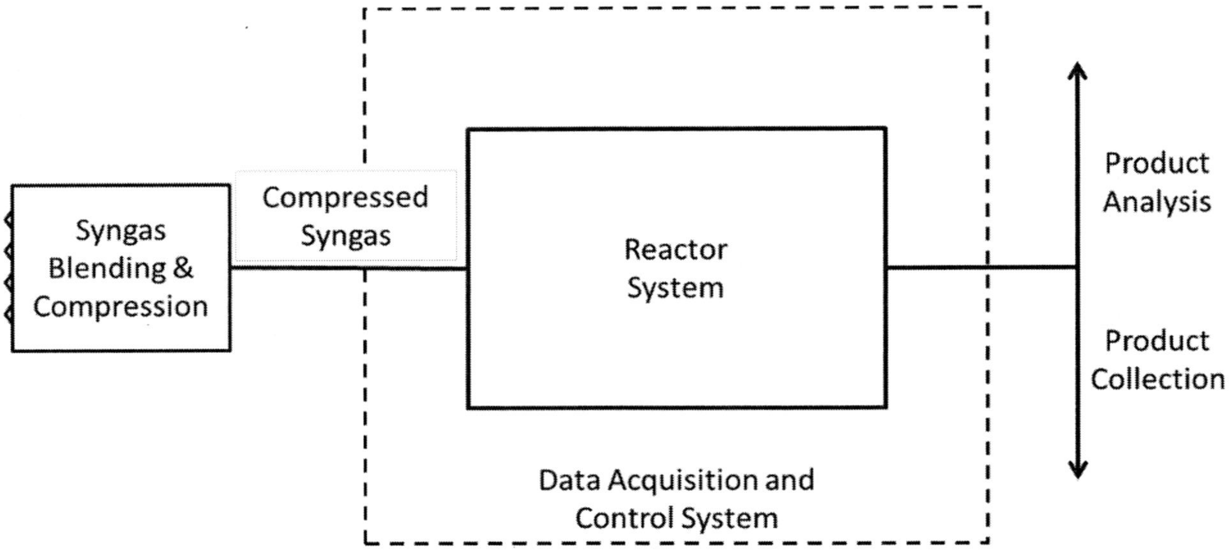

Figure 1. Design of the Mini-pilot of Single-stage Syngas-to-DME Process.

Acknowledgment

This research work was partially supported through research contracts from the Korea Institute of Science and Technology (KIST). The authors are grateful for their support.

References

[1] S. Lee (PI) and M. R. Gogate, "Dimethyl Ether Synthesis Process", EPRI TR-100246 (Licensable Material), pp. 1-179, EPRI, Palo Alto, CA, February, 1992.

513058 Enhancing Lubricity of Gas-to-Liquid (GTL) Paraffinic Kerosene Using Additives and Optimizing the Fuel™s Physicochemical Properties

Abstract:

The aviation industry is growing strong with an average growth rate of 6.2% over the last decade. It is estimated that the demand for jet fuel to be doubled by the year 2040, thus increasing the greenhouse gas (GHG) emission significantly by 2050 [1]. Synthetic paraffinic kerosene (SPK) which is a product of Gas to Liquid Technology (GTL) is an ultra-clean fuel since it has low aromatics content and negligible quantities of sulfur compounds. As the SPK is environmentally benign, it has a good potential to replace conventional fuel Jet A1; however, SPK suffers from lower lubricity compared to its conventional counterpart. Lubricity is a measure of the fuelâ€™s ability to reduce friction and wear between the parts of the engine thereby making it a very important propriety for extending engine life. Currently, the industry is restricted to use a maximum of 50:50=SPK: JetA1 as mentioned in ASTM-7566 [2].

To improve the lubricity of a GTL kerosene three additives have been selected: linoleic acid, ethyl oleate, and quinoline. An optimization of the physicochemical properties concerning the additives concentrations has been conducted. Statistical regression method has been used to carry out the optimization for each physicochemical property. Each additive is blended with SPK at five different concentrations, and the lubricity of blends has been tested using test method D5001. Samples that meets the industry specifications (ASTM-1655/7566) for lubricity were further investigated to study the effect of the additive on density, flash point, freezing point, viscosity, and calorific value and later compare the results with pure SPK [3]. Every property has been determined experimentally following respective test method as mentioned in (ASTM-1655/7566) standards. Through this novel fuel blends have been identified that either meet or exceed the industry specifications for aviation fuel. Finally, the optimized fuel blends have been combusted in an internal combustion engine to study their emissions and compare it to pure SPK and conventional fuel.

The result of this study will help identify potential fuel additives for SPK that enhance the lubricity without compromising other critical physicochemical properties of a jet fuel. Additionally, a combustion study will help understanding impact of such fuel blends on the environment. This study will also attempt to find optimum SPK blends that have potential to be utilized, SPK in engines of commercial flights without further blending it with conventional fuel.

Reference:

[1] U.S. Energy, â€œ International Energy Outlook 2017 Overview,â€ *Int. Energy Outlook,* vol. IEO2017, no 2017, p. 143,2017.

[2] I A. Al-Nuaimi, M. Bohra, M. Selam, H.A. Choudhury, M. M. El-Halwagi, N.O.Elbashir, Optimization of the aromatic/paraffinic composition of synthetic jet fuels, *Chem. Eng. Technol.*, Vol.39, p 2217, 2016

[3] E.E. Elmalik, B. Raza, S. Warrang, H. Ramadhan, E. Alborzi, N.O. Elbashir, Role of hydrocarbon building blocks on gas to liquid derived synthetic jet fuel characteristics, *Ind. Eng. Chem.*, vol. 53, p 1856, 2014

513060 Effect of Space Velocity and Long Residence Time to Fischer Tropsch Products and Catalysts Reduced with Different Gases

Abstract:

In this paper, we carried out Fischer-Tropsch synthesis (FTS) over an industrial iron-based catalyst ($FeCuKSiO_2$) under unusual laboratory operating conditions, involving an extremely low reaction pressure (1.85 bar absolute) and prolonged reaction times on stream (TOS) that are representative of industrial practice (5500 h). Three different reducing gases (syngas $H_2/CO = 2$, H_2 and CO) were used to activate the catalyst. FTS was conducted under similar conditions (gas hourly space velocity {GHSV} at 2592 h^{-1}, 1.85 bar (abs) and temperature 533 K) under each reducing gas. The evolution of the morphology and microstructures of the spent catalysts that had been differently reduced were characterized by HRTEM and XRD. Running FTS for long TOS results in a change in the product distribution due to catalyst speciation as deactivation occurs. Magnetite is the dominant phase in a deactivated catalyst, and graphitic carbon is found on the catalyst surface. The deposited carbon tends to be a hindrance. Diffusion concert can be used to explain the trend observed in this work. The catalyst speciation that could be observed as the catalyst deactivated with TOS resulted in the formation of magnetite, which is a water gas shift (WGS)-active catalyst. FTS reactions are accompanied by the formation of carbonaceous by-products (like coke), which deactivate the catalyst. GHSV is preferential for an increase in light FTS products, while the opposite trend is to be seen in heavy products. From the observed results, it can be concluded that at very long TOS a trend inversion in terms of product selectivity occurs if the GHSV is changed. An analysis of the link between GHSV and product distribution has also been given using diffusion concepts.

512595 Dynamic Simulation of Natural Gas Liquefaction with Helium Recovery

Abstract:

Natural gas is fastest growing energy economy. Its production and demands are increasing rapidly. Natural gas is transported as liquid over long distances. In 2016, natural gas liquefaction nameplate capacity reached 340 million tonnes per annum (MTPA). By January 2017, over 115 MTPA liquefaction capacity is under construction and 879 MTPA liquefaction capacity is proposed. Natural gas mainly contains methane. Lighter hydrocarbons and other components such as acid gas, mercury, carbon dioxide, water, nitrogen, helium are also present in natural gas. Natural gas pre-treatment is necessary to meet specifications set by standards and customers. The impurities are removed from natural gas before and during liquefaction process.

Helium is one of the precious components with a range of uses. It has limited reserves. The increasing demand and price of helium makes it more important to recover from natural gas. Natural gas reserves are the main resource of helium. Major portions of available helium exist in the United States. Helium can be recovered by processes such as cryogenic separation, adsorption, membrane separation. In this paper, steady-state and dynamic simulations are employed to study the helium recovery together with the natural gas liquefaction process. Sensitivity analysis of the helium recovery performance based on varying operating conditions such as feed composition, flow rate, and natural gas liquefaction temperature are also studied. The optimum operating conditions are suggested for the minimum energy consumption and maximum recovery of grade A helium. The optimization study may provide significant benefits in both LNG and helium productions in North America, where abandon reserves of natural gas are being monetized.

Experimental Study and CFD Simulation of the Multiphase Flow Conditions Encountered in a Novel Down-Flow Bubble Colum

18AIChE Spring Meeting · 14th Global Congress on Process Safety · Orlando Florida 2018

Mutharasu L.C.[1*], Dinesh V. Kalaga[2], Mayur Sathe[1], Derek Griffin[3], Xueliang Li[3], Masahiro Kawaji[2], Krishnaswamy Nandakumar[1] and J. B. Joshi[4]

1Louisiana State University, USA.

2The CUNY Energy Institute at the City College of New York, New York, NY, USA10031.

3LanzaTech, Skokie, IL, USA-6007.

4Homi Bhabha National Institute, Anushaktinagar, Mumbai-400094, India.

1

Motivation

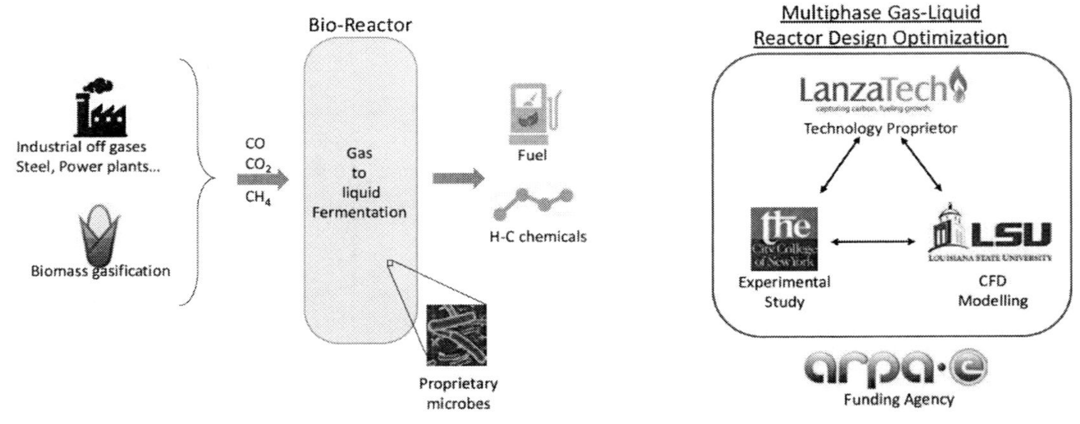

2

Why Down-flow Bubble columns ?

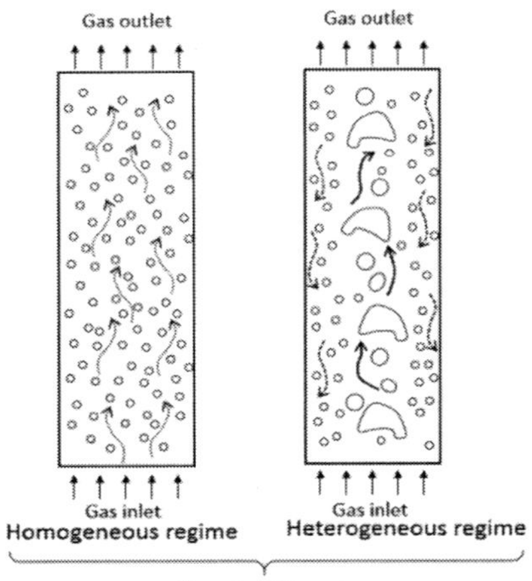

Up-flow Bubble column

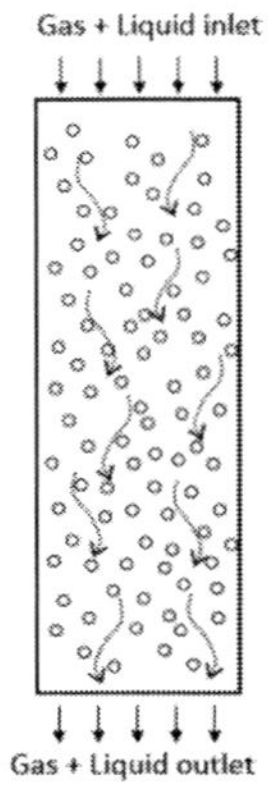

Down-flow Bubble column

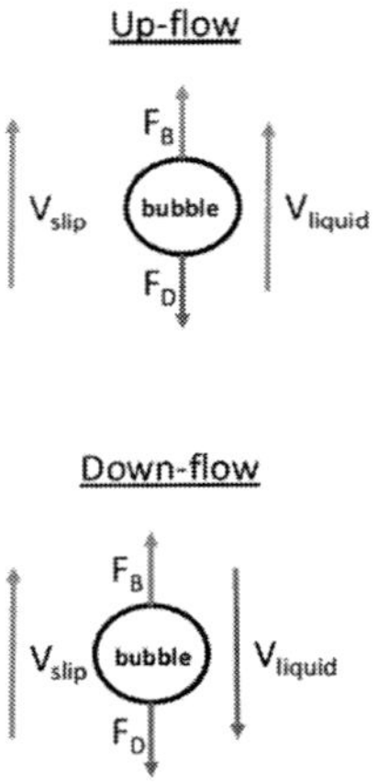

3

Novel Features of the Bubble column

Conventional Plunging jet Down flow Bubble column

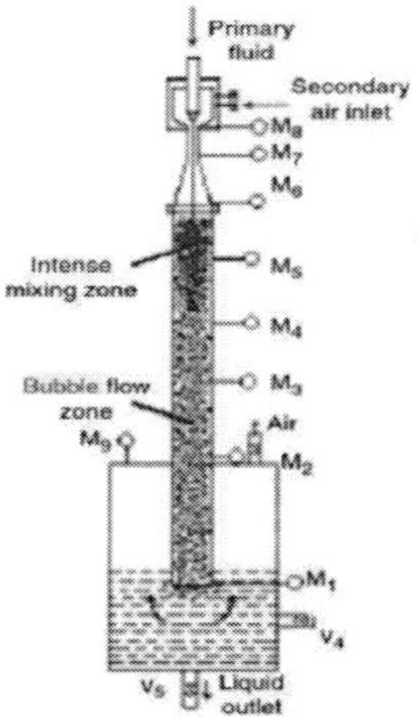

Conventional Plunging Jet Down-flow Bubble column (Majumder et al, 2005)

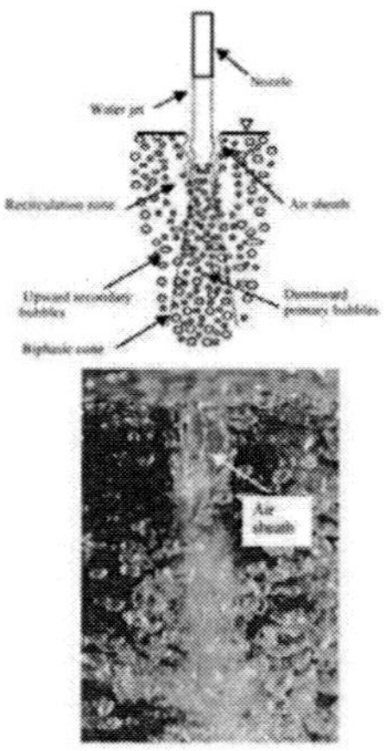

(Roy et al, 2013)

Novel Down flow Bubble column

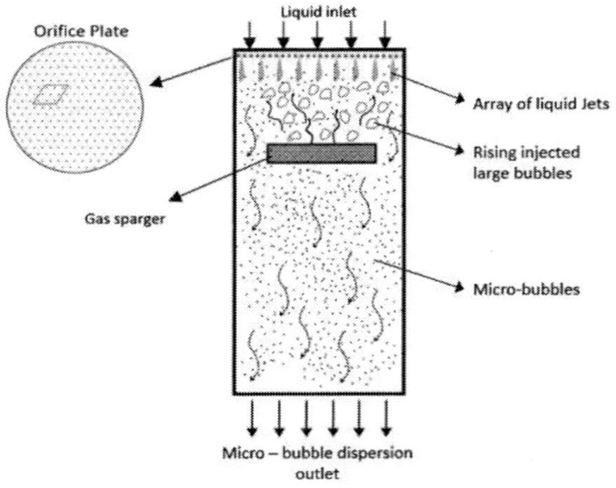

4

Experimental Setup

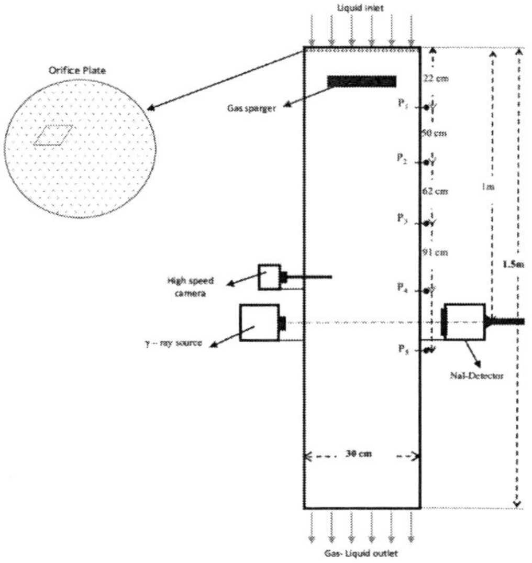

1. Inverse Bubbly flow with large cross-section.
2. Micro-bubble generation by impinging liquid jets
3. Gas injection.

5

Experimental Setup

Bubble column

8 arm gas sparger

Gamma Ray Densitometry setup

Measurement Techniques

1. **Phase isolation technique** (Volumetric gas holdup)
2. **Pressure transducer** (axial variation of gas holdup)
3. **Gamma Ray Densitometry** (Radial Variation of gas holdup)
4. **High Speed Camera & Borescope** (bubble size)

High speed camera images of Micro-bubbles

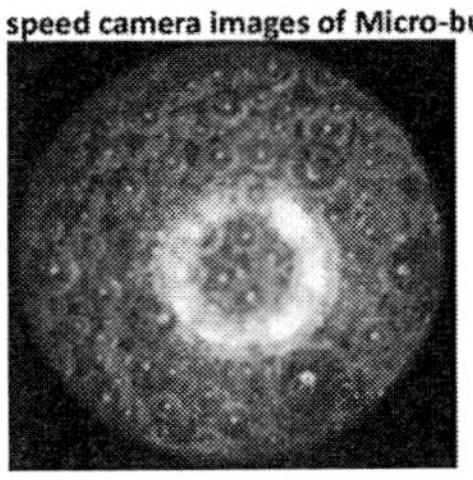

6

Experimental Results

Volumetric Gas holdup from phase isolation technique

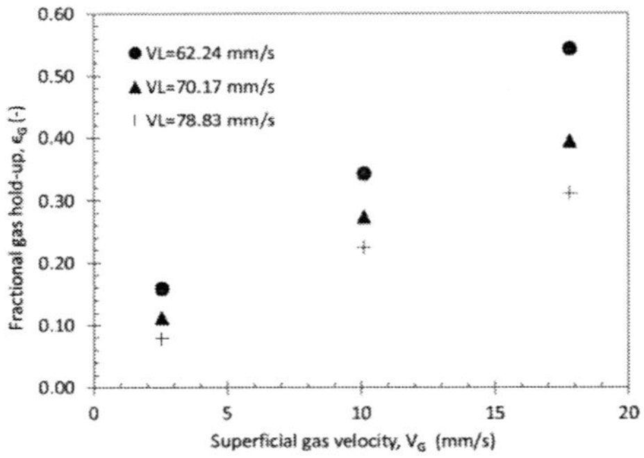

Axial variation of Gas holdup from pressure transducers

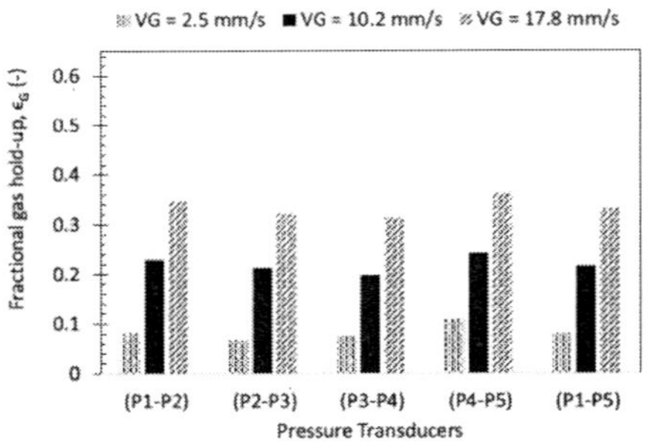

7

Experimental Results

Bubble-size-distribution from Camera and Borescope assembly
(100 cm below the orifice plate at r/R =0)

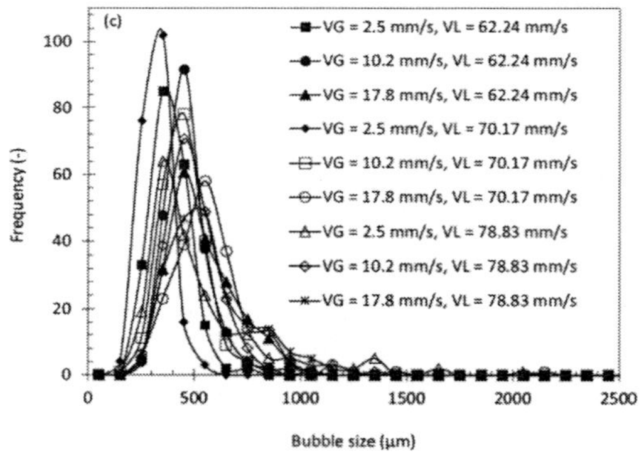

Radial variation of Mean-bubble –size
(100 cm below the orifice plate)

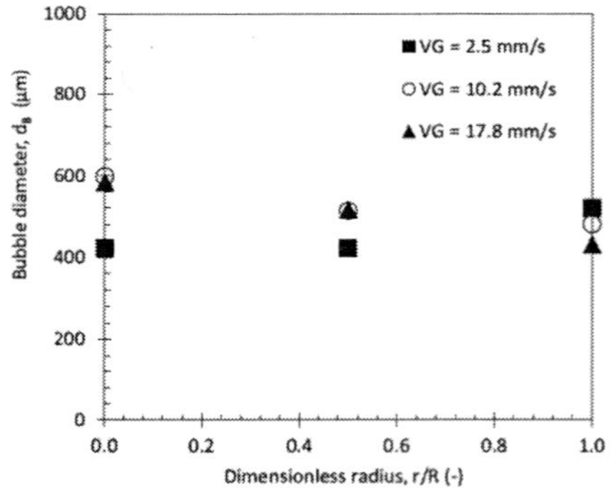

8

Multiphase CFD Models

EULER – EULER MODEL

EULER – LAGRANGE MODEL

VOLUME OF FLUID

LEVEL SET

DIRECT NUMERICAL SIMULATION

Euler – Euler model provides necessary flow information with least computational requirement.

9

Eulerian-Eulerian framework

- Continuity Equations:

$$\frac{\partial}{\partial t}(\alpha_k \rho_k) + \nabla.(\alpha_k \rho_k \vec{v}_k) = 0$$

- Fluid-Fluid Momentum Equations:

$$\frac{\partial}{\partial t}(\alpha_k \rho_k \vec{v}_k) + \nabla.(\alpha_k \rho_k \vec{v}_k \vec{v}_k) = -\alpha_k \nabla p + \nabla.\bar{\bar{\tau}}_k + \alpha_k \rho_k \vec{g} + \left(\vec{F}_{drag,k} + \vec{F}_{lift,k} + \vec{F}_{wl,k} + \vec{F}_{vm,k} + \vec{F}_{td,k}\right)$$

Where, $\bar{\bar{\tau}}_k = \alpha_k \mu_k \left[(\nabla \vec{v}_k + \nabla \vec{v}_k^T) - \frac{2}{3}(\nabla.\vec{v}_k)I\right]$

10

Turbulence Modelling

Mixture K- epsilon model

K - equation

$$\frac{\partial}{\partial t}(\rho_m k) + \nabla.(\rho_m k \vec{v}_m) = \nabla.\left(\frac{\mu_{t,m}}{\sigma_k}\nabla k\right) + G_{k_m} - \rho_m \varepsilon$$

Epsilon - equation

$$\frac{\partial}{\partial t}(\rho_m \varepsilon) + \nabla.(\rho_m \varepsilon \vec{v}_m) = \nabla.\left(\frac{\mu_{t,m}}{\sigma_\varepsilon}\nabla \varepsilon\right) + \frac{\varepsilon}{k}(C_{1\varepsilon}G_{k_m} - C_{2\epsilon}\rho_m \varepsilon)$$

Where:

$$\rho_m = \sum_{k=1}^{N} \alpha_k \rho_k$$

$$\vec{v}_m = \frac{\sum_{k=1}^{N} \alpha_k \rho_k \vec{v}_k}{\sum_{k=1}^{N} \alpha_k \rho_k}$$

$$\mu_{t,m} = \rho_m C_\mu * \frac{k^2}{\epsilon}$$

$$G_{k,m} = \mu_{t,m}(\nabla \vec{v}_m + (\vec{v}_m)^T): \nabla \vec{v}_m$$

11

Source point modelling

Modelling of Jets as source points

Mesh

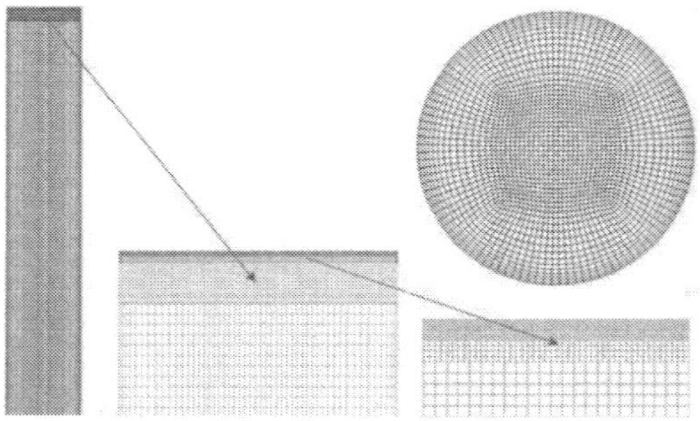

Cells marked for Source point modelling

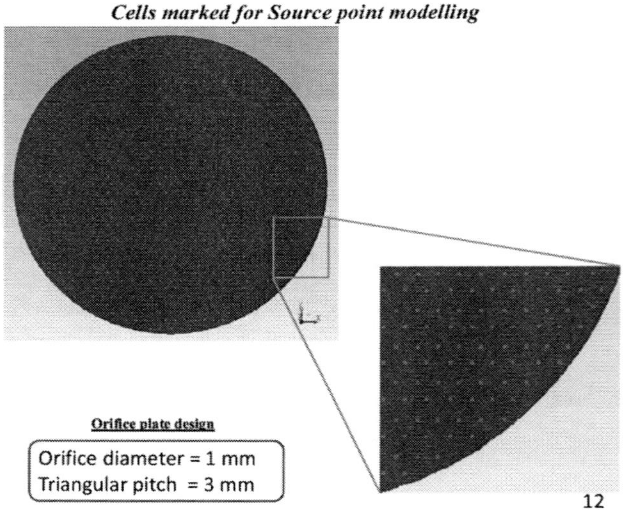

Orifice plate design

Orifice diameter = 1 mm
Triangular pitch = 3 mm

12

Source point modelling

Modelling of Gas sparger as source points

Cells marked for Source point modelling

8 arm gas sparger

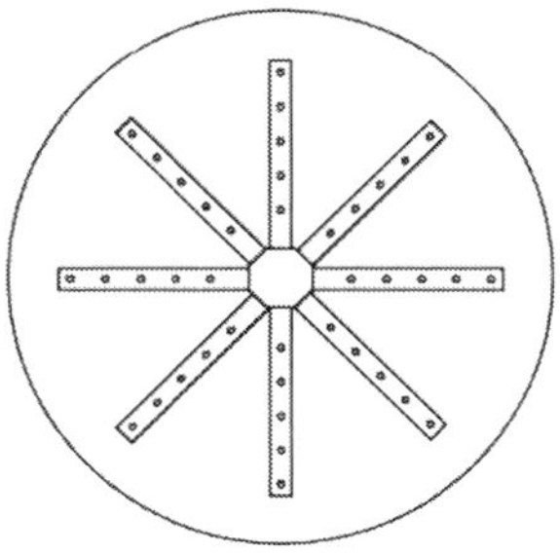

Large Bubble to Micro-bubble breakage modelling

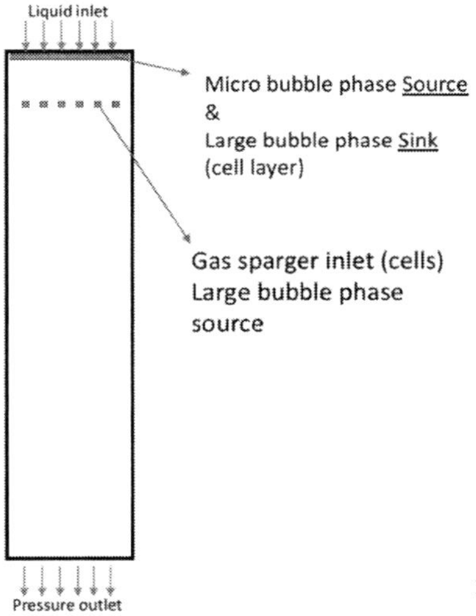

13

Drag modification from Force Balance

Slip Velocity (Simple Force Balance)

$$F_{drag} = D_f \frac{A_i}{8} C_D \rho_l |v_L - v_G| \, (v_L - v_G)$$

$$F_B = \epsilon_G \, g \, (\rho_L - \rho_G)$$

$$V_{s_{fb}} = \sqrt{\frac{4d_B}{3D_f C_D} \frac{\epsilon_G}{(\epsilon_L - \epsilon_G)} \left(\frac{\rho_L - \rho_G}{\rho_L}\right) g}$$

$$D_f = (1 + \alpha) * (1 - \epsilon_G)^n$$

Experimental Average Slip Velocity

VG (mm/s)	VL (mm/s)	Gas holdup (-)
2.54	62.24	0.96
10.1	62.24	0.84
17.8	62.24	0.71
2.54	70.17	0.96
10.1	70.17	0.86
17.8	70.17	0.75
2.54	78.83	0.97
10.1	78.83	0.87
17.8	78.83	0.77

$$V_{s_{exp}} = \frac{V_L}{\epsilon_L} - \frac{V_G}{\epsilon_G}$$

14

Drag modification from force balance

Investigators	Drag Coefficient Expression
Schiller and Naumann (1935)	$C_D = \begin{cases} \frac{24(1+0.15Re_g^{0.687})}{Re_g}, & Re_g < 1000 \\ 0.44, & Re_g \geq 1000 \end{cases}$
Dalla Ville (1948)	$C_D = \left(0.63 + \frac{4.8}{\sqrt{Re_g}}\right)^2$
Clift et al. (2005)	$C_D = \begin{cases} \frac{29.1667}{Re_g} - \frac{3.8889}{Re_g^2} + 1.222, & 1 < Re_g < 10 \\ \left(\frac{24}{Re_g}\right)(1+0.15Re_g^{0.687}), & 10 < Re_g \leq 200 \end{cases}$
Ma and Ahmadi (1990)	$C_D = \frac{24}{Re_g}(1+0.1Re_g^{0.75})$
Mei et al. (1994)	$C_D = \frac{16}{Re_g}\left\{1 + \left[\frac{8}{Re_g} + \frac{1}{2}(1+3.315Re_g^{-0.5})\right]^{-1}\right\}$
Laín et al. (2002)	$C_D = \begin{cases} \frac{16}{Re_g}, & 1 \leq Re_g \\ \frac{14.9}{Re_g^{0.78}}, & 1.5 \leq Re_g < 80 \\ \frac{48}{Re_g}\left(1 - 2.\frac{21}{\sqrt{Re_g}}\right) + 1.86 \cdot 10^{-15}, & 80 \leq Re_g < 1500 \\ 2.61, & 1500 \leq Re_g \end{cases}$
Tomiyama et al. (1998)	$C_D = max\left\{min\left[\frac{24}{Re_g}(1+0.15Re_g^{0.687}), \frac{72}{Re_g}\right], \frac{8}{3}\frac{E_o}{E_o+4}\right\}$
Tomiyama (2004)	$C_D = \frac{8}{3}Eo\frac{(1-E^2)}{E^{2/3}E_o + 16(1-E^2)E^{4/3}}F(E)^{-2}$ $F(E) = \frac{sin^{-1}\sqrt{1-E^2} - E\sqrt{1-E^2}}{1-E^2}, \quad E = \frac{1}{1+0.163E_o^{0.757}}$
Zhang and VanderHeyden (2002)	$C_D = 0.4 + \frac{24}{Re_g} + \frac{6}{1+\sqrt{Re_g}}$

With no Drag modification

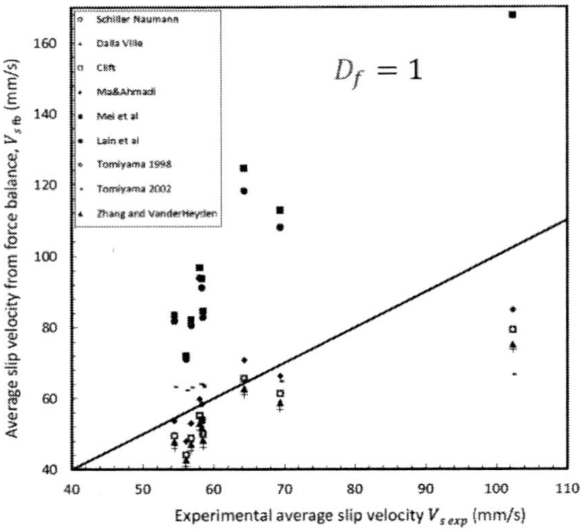

Drag modification from force balance

With optimum D_f values

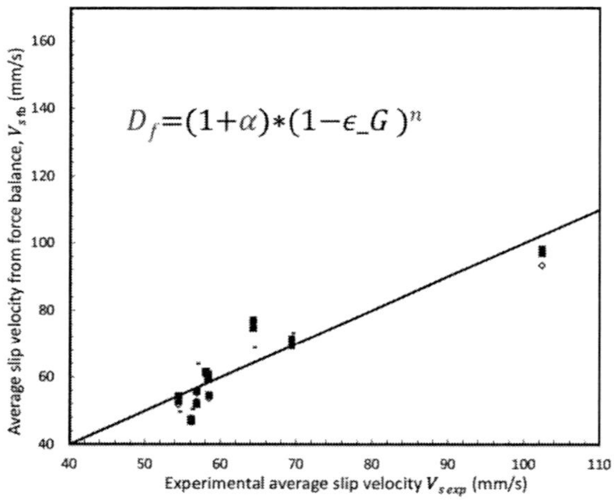

Optimized values of 'α' and 'n' for the different drag correlations obtained from the Force Balance Method.

Drag model	α	n
Schiller and Naumann (1935)	-0.07	-0.70
Dalla Ville (1948)	-0.16	-0.69
Clift et al. (2005)	-0.07	-0.70
Ma and Ahmadi (1990)	0.05	-0.68
Mei et al. (1994)	0.64	-1.14
Laín et al. (2002)	0.64	-1.06
Tomiyama et al. (1998)	-0.03	-0.63
Tomiyama (2004)	0.88	0.77
Zhang and VanderHeyden (2002)	-0.11	-0.63

Drag modification from force balance

Comparison between various formulations of the drag modification factor D_f

Reference	C_D/C_{D_∞}	$d_B\ (mm)$	ϵ_G range
Davidson and Harrison (1966)	$(1-\epsilon_G)^2$	3.5	0-0.2
Griffith and Wallis (1961)	$(1-\epsilon_G)^{-2}$	-	-
Richardson and Zaki (1954)	$(1-\epsilon_G)^{-2.78}$	0.1	0.4-0.9
Marrucci (1965)	$\frac{\left(1-\epsilon_G^{5/3}\right)^2}{(1-\epsilon_G)^4}$	-	-
Lockett and Kirkpatric (1975)	$\frac{\left(1+2.55\,\epsilon_G^3\right)^{-2}}{(1-\epsilon_G)^{2.78}}$	5	0-0.66
Joshi and Lali (1984)	$1+\frac{(1.5\,\epsilon_G)^2}{g d_B/3V_{b\infty}^2}$	3-6	0-0.3
Garnier et al. (2002)	$\left(1-\epsilon_G^{1/3}\right)^{-2}$	3-4	0-0.35
Simonnet et al (2007)	$(1-\epsilon_G)\left[(1-\epsilon_G)^m+\left(4.8\frac{\epsilon_G}{1-\epsilon_G}\right)^m\right]^{-2/m}$	5-10	0.15-0.3

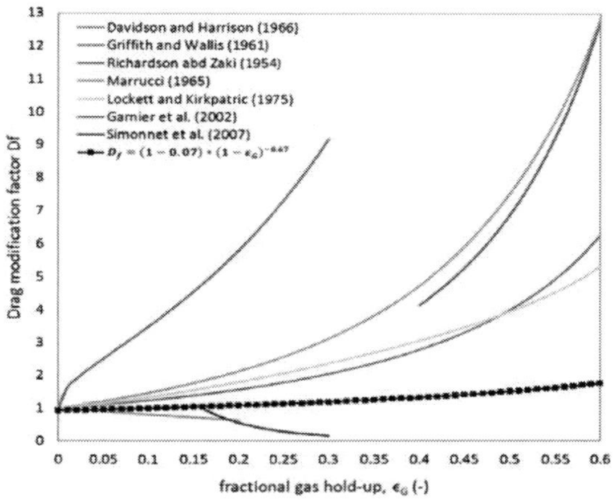

17

CFD Results

Contours for V_G=17.8 mm/s and V_L=62.2 mm/s.

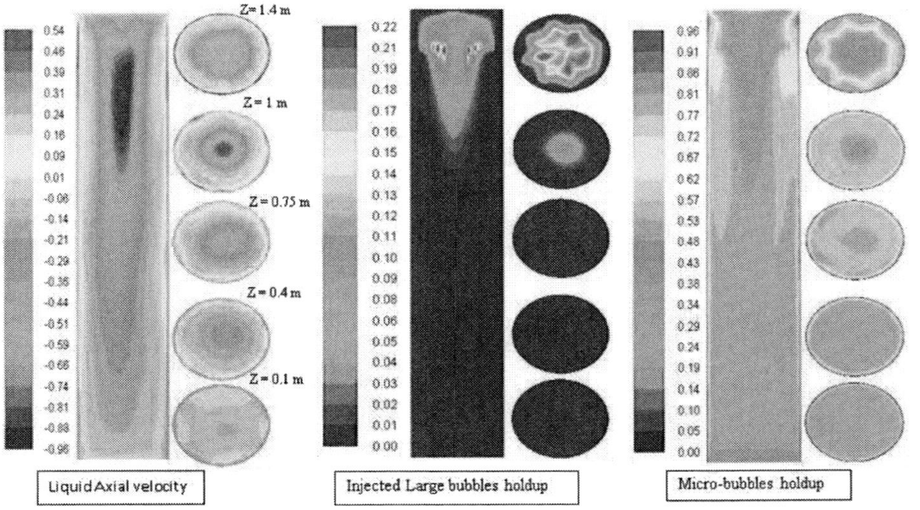

18

CFD Results (Radial profile of gas holdup)

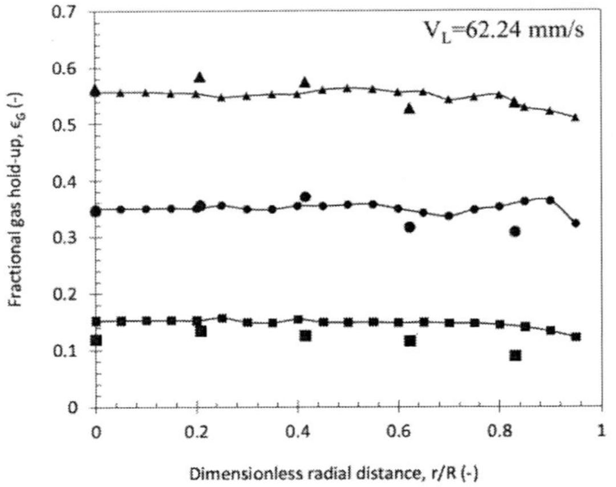

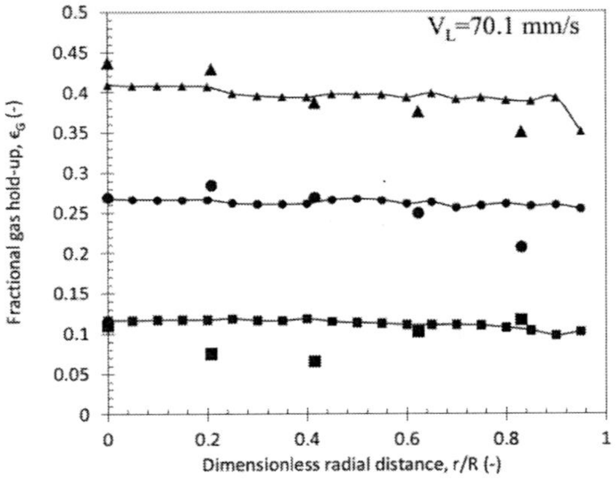

■ V_G = 2.5 mm/s (exp) **● V_G = 10.2 mm/s (exp)** **▲ V_G = 17.8 mm/s(exp)**

-■- V_G = 2.5 mm/s (CFD) **-●- V_G = 10.2 mm/s (CFD)** **-▲- V_G = 17.8 mm/s(CFD)**

19

CFD Results (Radial profile of gas holdup)

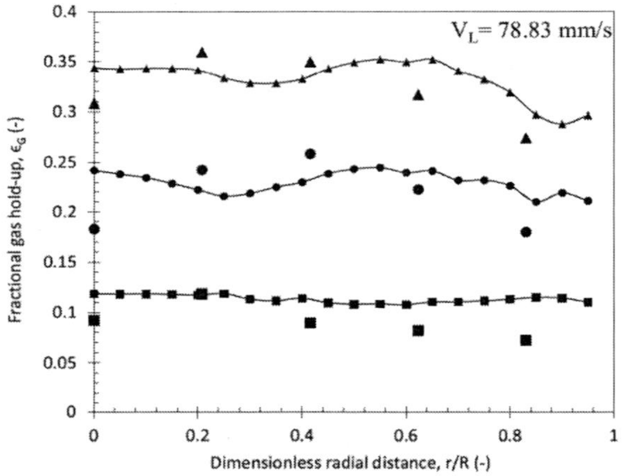

■ V_G = 2.5 mm/s (exp) **● V_G = 10.2 mm/s (exp)** **▲ V_G = 17.8 mm/s(exp).**

-■- V_G = 2.5 mm/s (CFD) **-●- V_G = 10.2 mm/s (CFD)** **-▲- V_G = 17.8 mm/s(CFD).**

20

CFD Results (Radial profile of Liquid velocity)

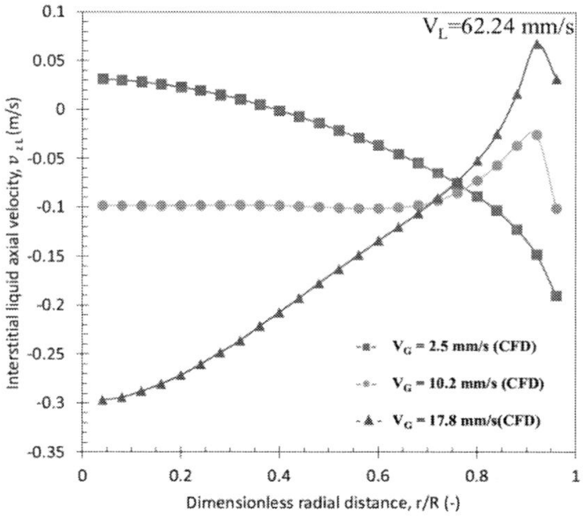

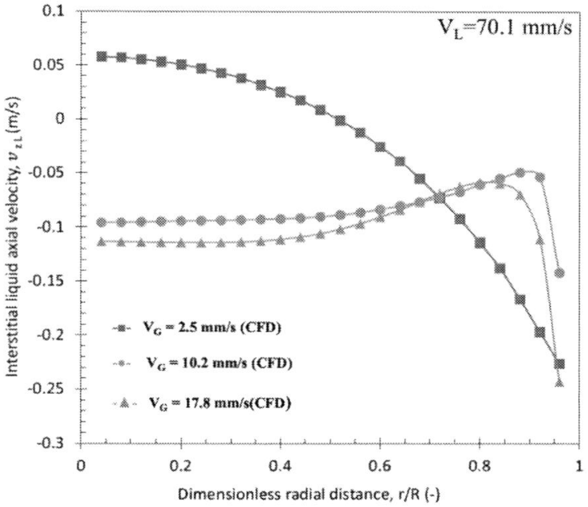

21

CFD Results (Radial profile of Liquid velocity)

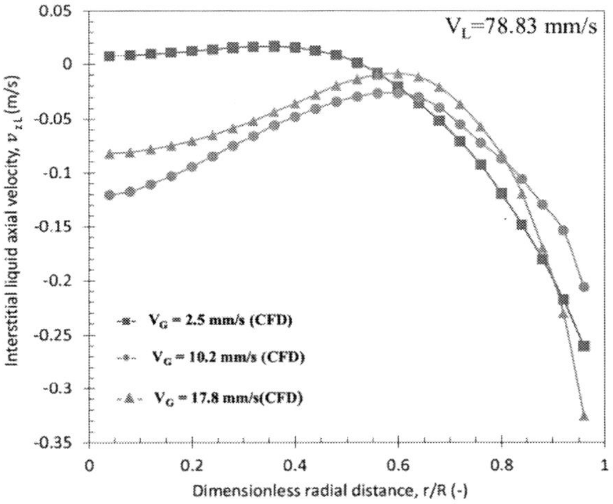

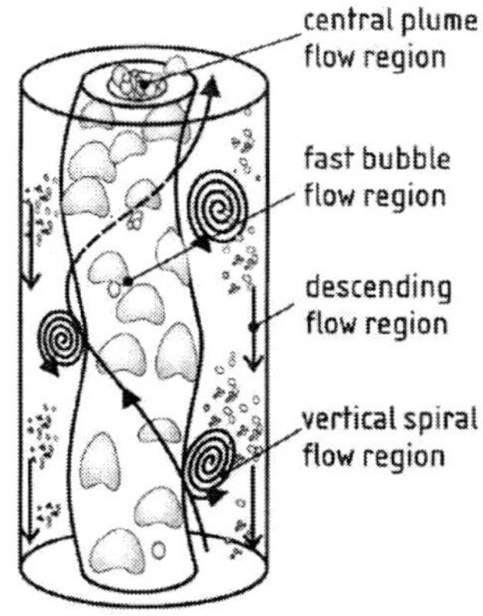

Illustration of the liquid phase flow circulations encountered in a conventional up-flow bubble column source: (Chen et al., 1994)

22

Conclusions

- Both Experimental data and CFD analysis, indicate a flat radial gas hold-up profile.
- The Drag modification factor $'D_f'$ takes into account both the swarm-effect (local gas holdup dependence) and also the effect of the surface-active additives on the gas-liquid interface.
- Simple Force Balance method was used for obtaining the optimum parameters of D_f
- CFD analysis, revealed that the flow is pure down-flow with no liquid circulations for most of the operating conditions, well within the operating window.
- The circulations observed at the edge of the operating window are not buoyancy driven.

23

Application of Dynamic Simulation for Relief System Design in LNG Facility

2018 AIChE Spring Meeting.
Topical 6: 18th Topical Conference on Gas Utilization
LNG Simulation, Operations and Control

Authors

Sameera Thakore - ssthakor@bechtel.com
Haribabu Chittibabu - hchittib@bechtel.com

Key words

LNG (Liquefied Natural Gas), HP (High Pressure), LP (Low Pressure), Steady State, Dynamic simulation, Flare, PSV (Pressure Safety Valves), Spec-Break.

1.0 Abstract

Emergency relief in the process industries aims to protect equipment, the environment and operating personnel from abnormal overpressure conditions. Appropriate determination of relief loads is important for the correct sizing of relief valves and flare headers, and for the selection of disposal media. In addition, during debottlenecking or revamping of process units, additional relief valves and/or modification of the relief system can be very costly and, in terms of construction, difficult to implement.

Application of dynamic modeling for relief system design can reduce the capital expenditure (CAPEX) without compromising on the safety. Dynamic simulation, utilizing rigorous mathematical models, has become an influential tool in process design, design validation, relief system design, control system verification, startup support and troubleshooting. Many recent developments in the simulation software and technology have led to the development of large scale dynamic models.

In the Inlet Facility of an LNG or Gas Plant, the feed gas or mixture of gas and condensate are received by means of high pressure pipelines. The condensate or liquid slugs, if any, are separated in the Inlet Slug Catcher / Inlet Separator and feed gas may split into multiple trains for processing. The Condensate Feed Drum is also part of inlet facility for an LNG or Gas Plant that receives and separates front end liquids from the Inlet Slug Catcher / Inlet Separator. The operating pressures of Inlet Slug Catcher / Inlet Separator are significantly higher than the design pressure of Condensate Feed Drum, thereby creating HP/LP interfaces, and a risk of overpressuring the Condensate Feed Drum.

There are many HP/LP interfaces in a gas processing facility. Typically, the governing overpressure scenario for the vessel operating at low pressure occurs when the level control valve of the HP upstream vessel fails in the open position. During this scenario, the different stages of relief may occur as follows:

1. Initial liquid / two-phase relief: Liquid from the HP vessel flows through the control valve, and may flash at the lower pressure.
2. Liquid displacement: Once the liquid in the upstream vessel has emptied, vapor begins flowing through the line and upon reaching the control valve will expand, resulting in acceleration of the liquid in the downstream piping to the LP vessel.
3. Gas blowby: Vapor from the upstream HP vessel flows through the control valve to the downstream LP vessel.

Conventional methodology for determining the relief load does not consider any hydraulic losses in the system, resulting in overestimation of the required relief rate. In general, if the liquid volume in the upstream vessel exceeds the available vapor/liquid disengagement volume in the downstream vessel, then the liquid displacement portion of the relief results in the greatest required relief due to the expansion of vapor across the control valve. If the liquid volume in the upstream vessel does not exceed the available vapor/liquid disengagement space in the downstream vessel, then the greatest required relief occurs during gas blowby.

Applying dynamic simulation, the relief load is limited based on the resistance in the system.

This paper considers different methods for estimating relief load for a HP/LP interface system, compares the results from both steady state and dynamic methods relief rate determination, and discusses the strengths and weaknesses of each method. Furthermore, it describes how dynamic simulation was applied to an existing LNG facility in reducing the capital expenditure (CAPEX) and reducing the turnaround time for the modification.

2.0 LNG Plant Inlet Facility and Stabilizer Feed Drum

The Stabilizer Feed Drum is a three-phase separator that receives and separates front end liquids from Slug Catchers, filters and KO drums in the front end treatment areas. The operating pressures of the source equipment are significantly greater than the design pressure of the Stabilizer Feed Drum, creating HP/LP interfaces and a risk of overpressure the Stabilizer Feed Drum.

Figure 1 provides an overall system sketch.

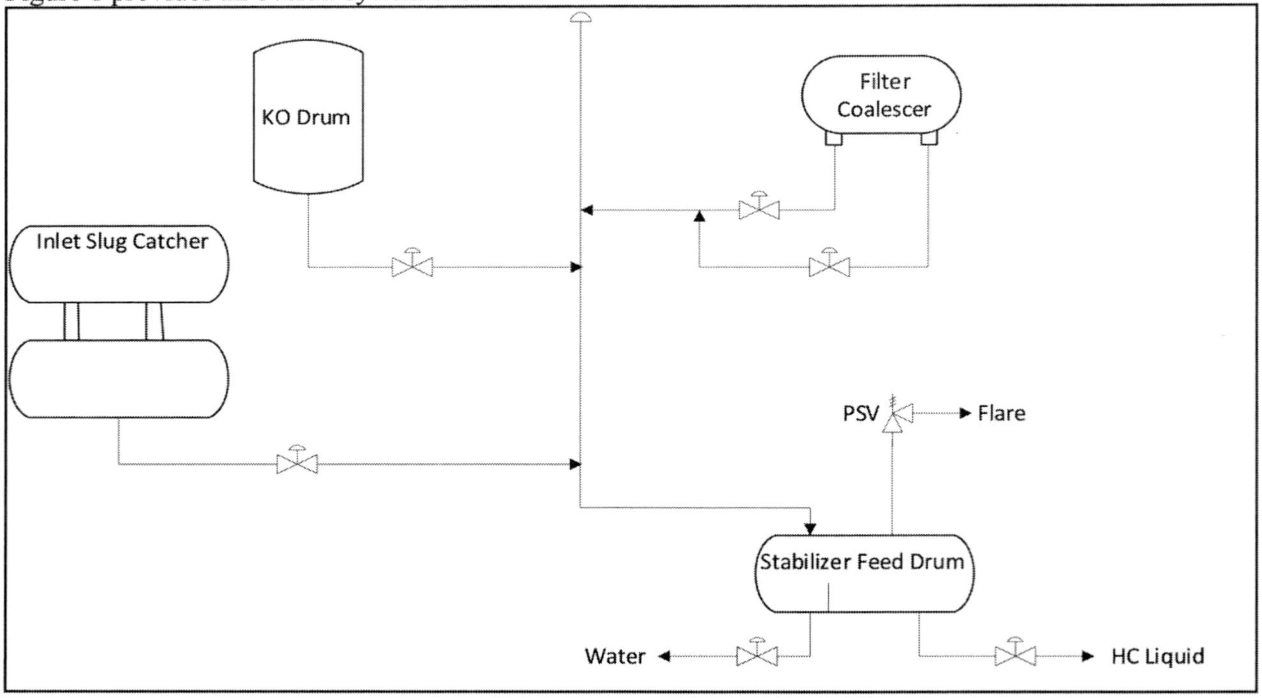

Figure 1. System Sketch

3.0 Relief Analysis

There are many HP/LP interfaces in an LNG facility. Typically, the governing relief scenario for the low pressure vessel is a level control valve on the liquid outlet of a high pressure upstream vessel failing wide open. Often, the volume of liquid in the upstream vessel exceeds the available vapor space in the downstream vessel. For the Stabilizer Feed Drum example, during this scenario, relief may occur in the following stages:

Upstream Vessel Contains Liquid Hydrocarbons

- The liquid from the upstream vessel flashes across the control valve and enters the Stabilizer Feed Drum as a two-phase fluid. Initially, there is disengagement space in the downstream vessel. As the event progresses, the liquid level increases. The entering gas will entrain sufficient liquid to occupy the disengagement space. The relief fluid is two-phase.

- After the liquid level in the upstream vessel is lost, the gas blowby event starts. Gas flows through the line and control valve and pushes the two-phase fluid, which remains downstream in the line to the Stabilizer Feed Drum. Relief continues to be two-phase.

- Gas blowby from the upstream vessel through to the Stabilizer Feed Drum. After the two-phase fluid has been completely pushed to the downstream vessel, the gas flows through.

Upstream Vessel Contains Liquid Water

- Initial relief is water. Water from the upstream vessel flows through the control valve to the downstream vessel, causing it to become liquid full. Initial relief is water.

- Vapor expansion occurs. Vapor flows through the line and expands at the control valve discharge pressure, causing acceleration of incompressible liquid in the downstream liquid full piping to the downstream vessel.

- Gas blowby from the upstream vessel through to the Stabilizer Feed Drum. After the water has been completely pushed to the downstream vessel, the gas blows through.

For the Stabilizer Feed Drum system, the control valve on the Inlet Slug Catcher outlet line failing open will be the controlling scenario for the relief because this equipment often operates at the highest pressure and the control valve on the liquid outlet line is larger than those on the other HP/LP interfaces. An analysis was performed on the control valve from the Inlet Slug Catcher failing wide open to compare results from the conventional method and dynamic simulation for determining the relief rate and PSV sizing.

3.1 Conventional Method

The conventional steady-state method for the determining the relief rate, which is conservative, is the following:

Relief Stage	Relief Fluid	Relief Rate Basis
Initial Relief	Two-Phase Hydrocarbon / Liquid Water	Wide-open flow through the control valve based on the upstream vessel at the design pressure and the downstream vessel at the relief pressure.
Vapor Displacement	Two-Phase Hydrocarbon / Liquid Water	Relief is two-phase fluid or liquid water at the volumetric rate of the vapor as it expands across the control valve. The vapor rate is determined based on the upstream vessel at the maximum operating pressure and the downstream vessel at the relief pressure.
Gas Blowby	Hydrocarbon Vapor	Wide-open flow through the control valve based on the upstream vessel at the design pressure and the downstream vessel at the relief pressure.

Table 1. Stages of Relief

3.2 Dynamic Simulation

Dynamic simulation may be used to evaluate this relief scenario and determine a more realistic relief rate.

The following are the model bases used for the dynamic simulation:

1. Upstream vessel is at the maximum operating pressure.
2. Destination pressure is set at relief pressure (Code allowable accumulation pressure).
3. Piping from the upstream vessel to the Stabilizer Feed Drum is modeled to account for hydraulic losses.
4. Upstream vessel initially contains liquid. The dynamic model considers the initial draining of liquid and the subsequent entrance of vapor into the line.

Figure 2 below provides VMGSim PFD of the dynamic model.

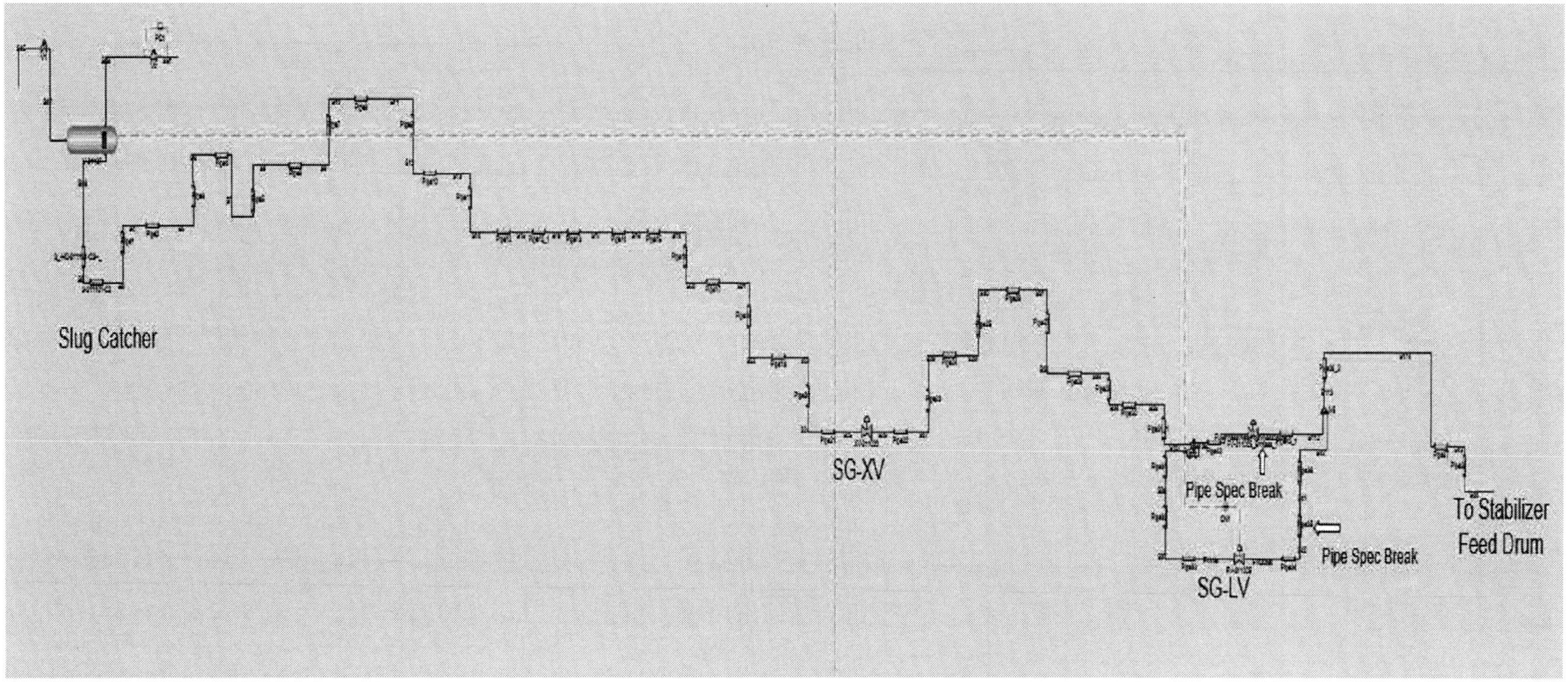

Figure 2: VMGSim PFD

Figures 3 and 4 that follow show the results of the dynamic evaluation for the relief rate.

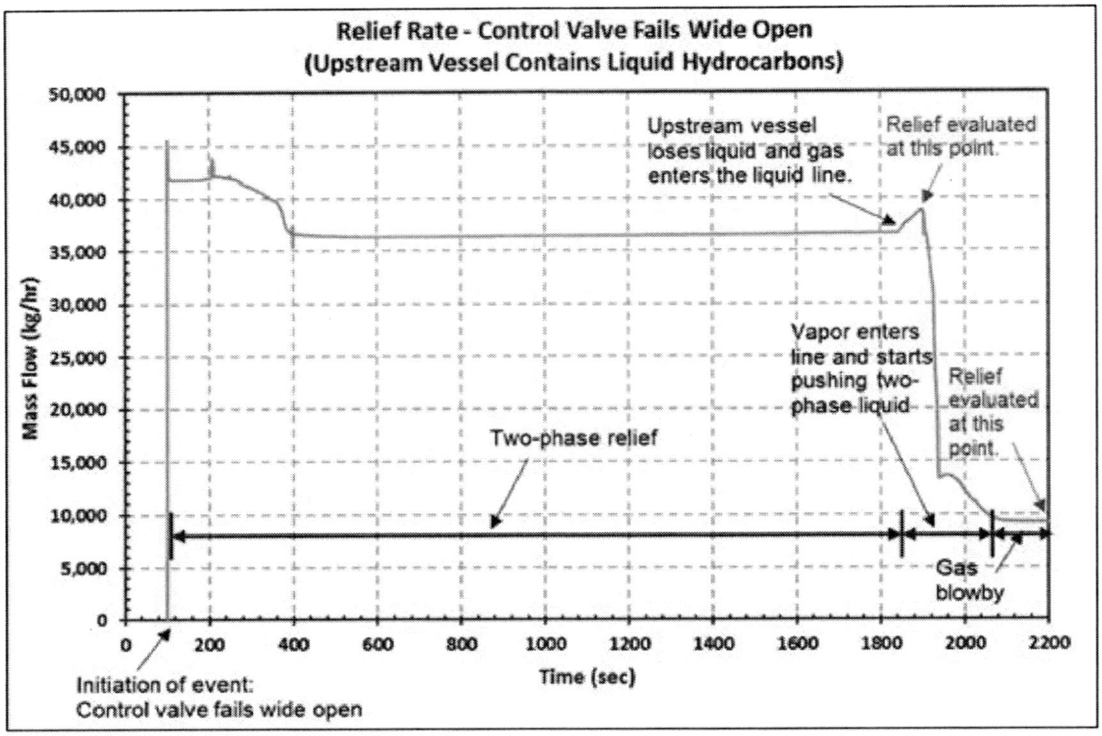

Figure 3: Dynamic Simulation Results – Upstream Vessel Contains Liquid Hydrocarbons

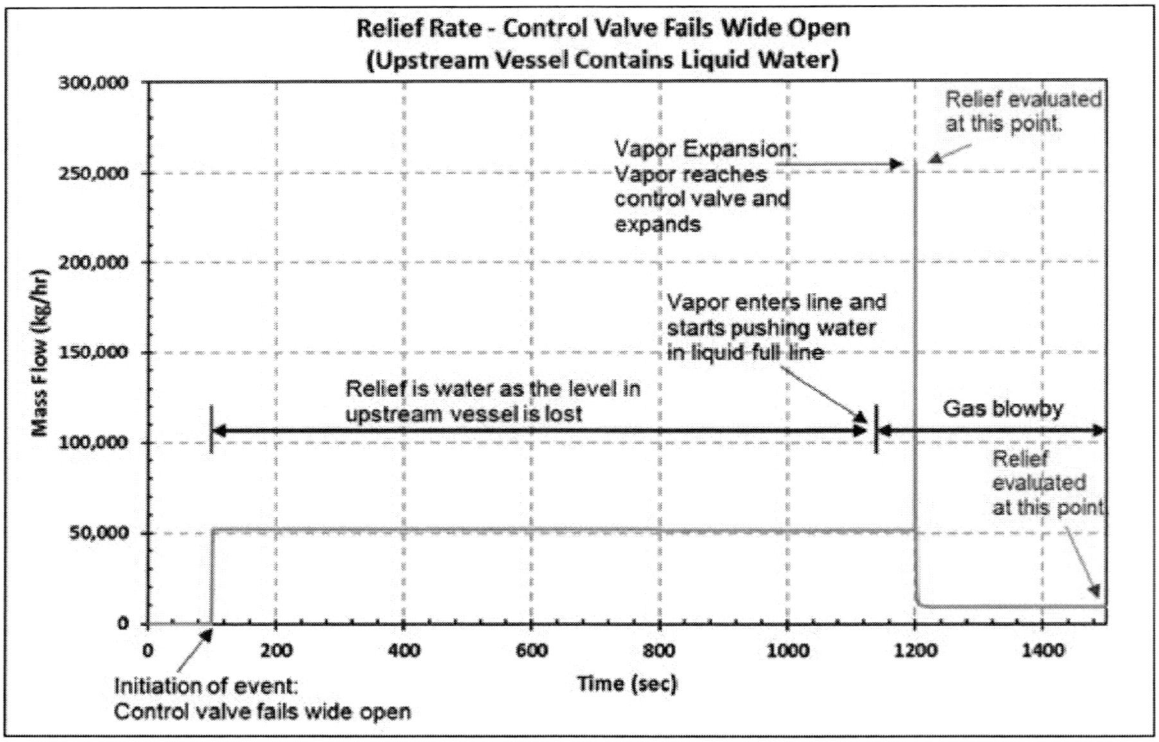

Figure 4: Dynamic Simulation Results – Upstream Vessel Contains Liquid Water

4.0 Conventional Method vs. Dynamic Method

Table 2 shows a comparison of the PSV orifice size required based on the conventional calculation and dynamic evaluation for the relief rates.

	PSV Orifice Size Required	
	Conventional Method	Dynamic Evaluation
Vapor Displacement – Upstream vessel contains liquid hydrocarbons	M	K
Vapor Displacement – Upstream vessel contains liquid water	P	L
Gas Blowby	J	J

Table 2. PSV Orifice Size Comparison

The conventional methodology for determining the relief load does not consider any hydraulic losses in the system, resulting in overestimation of the required relief rate. In general, if the liquid volume in the upstream vessel exceeds the available vapor/liquid disengagement volume in the downstream vessel, then the liquid displacement portion of the relief results in the greatest required relief due to the expansion of vapor across the control valve. If the liquid volume in the upstream vessel does not exceed the available vapor/liquid disengagement space in the downstream vessel, then the greatest required relief occurs during gas blowby.

For the gas blowby, hydraulic losses may have minimal impact on the relief rate determination and the conventional methodology provides an adequate estimation of the relief rate.

However, for the vapor displacement portion of the relief, the conventional method overestimates the relief load required, resulting in an oversized PSV. Potential consequences include:

- Increased cost of PSV
- Oversized inlet and outlet laterals (greater piping costs)
- PSV chattering

The dynamic evaluation for the relief load provides a more realistic estimation of the relief required by accounting for the resistance in the system.

5.0 Rerating of Piping Components

Dynamic simulation of the control valve failing open scenario also provides useful information regarding the backpressures in the piping system. During this scenario, the peak backpressure occurs during the vapor expansion phenomena.

The location of the control valve governs the backpressure on the HP/LP interface (spec break) when the PSV on Stabilizer Feed Drum is relieving.

If the control valve is close to the HP source, then the backpressure at the spec break is greater. If the control valve is close to the LP destination vessel, then the backpressure at the spec break is lower. Figure 5 shows a typical layout of Slug Catcher and Dehydration Unit Filter Coalescer level control valves with respect to Stabilizer Feed Drum.

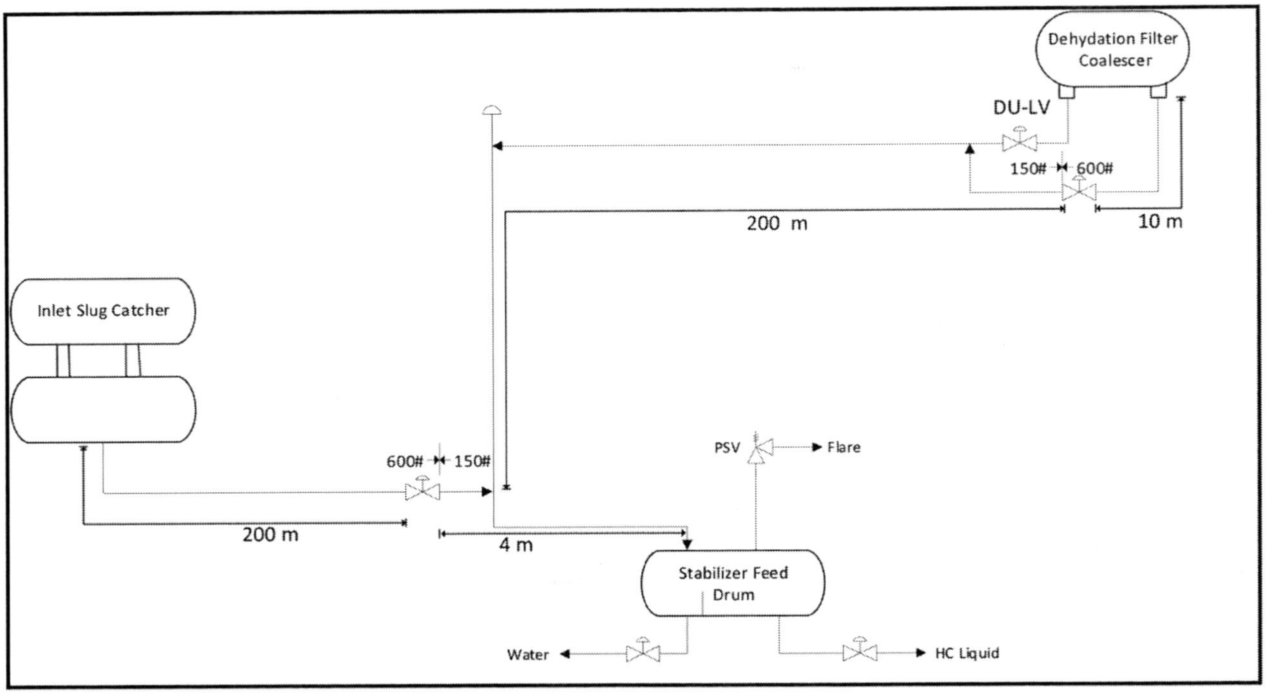

Figure 5. Layout of Inlet Facility and Dehydration Unit

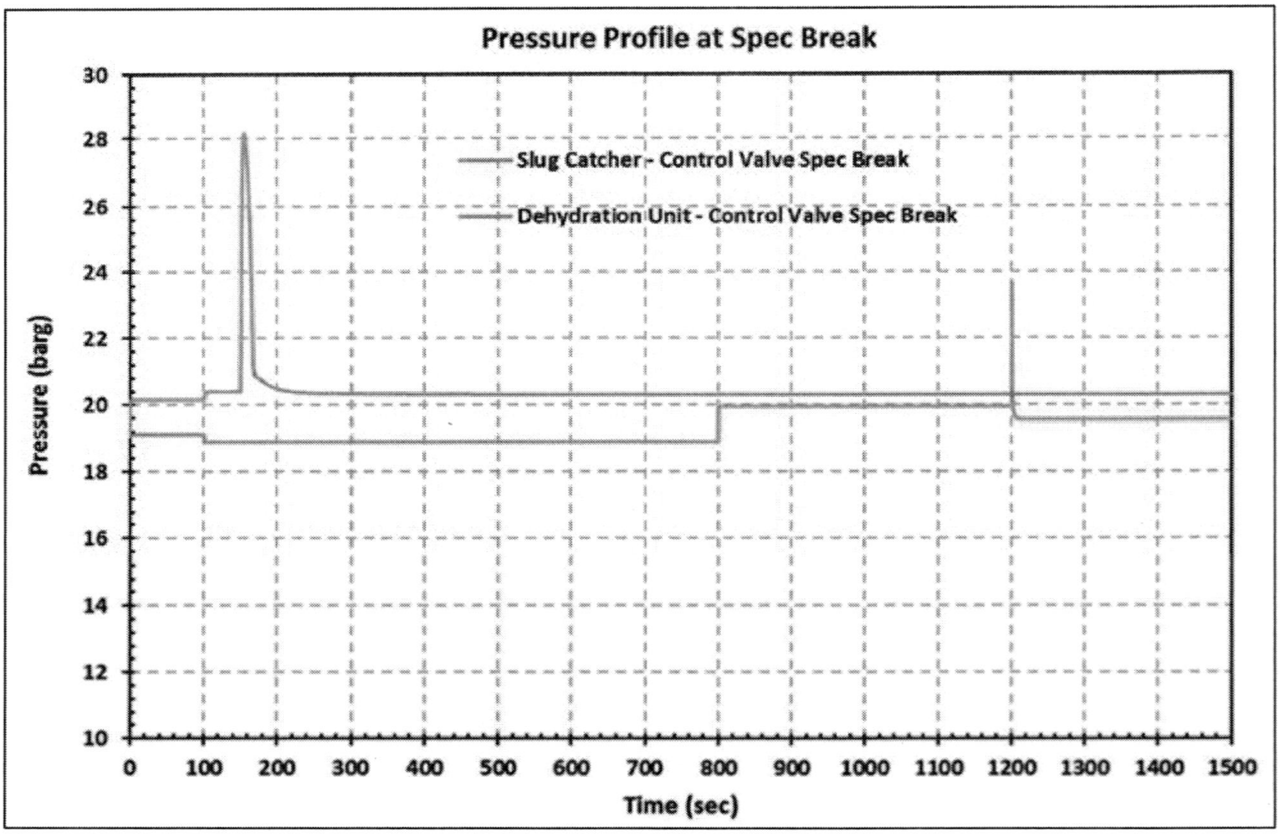

Figure 6. Pressure Profile at the Spec Break during Vapor Expansion

Figure 6 shows the pressure spike at the HP/LP interface (the spec break) during vapor expansion. The backpressure on the Dehydration Unit Filter Coalescer control valve outlet is greater than the backpressure on the Slug Catcher level control valve outlet.

Even though the Slug Catcher control valve failing open is the governing scenario for the Stabilizer Feed Drum PSV as shown in Figure 7, the Dehydration Unit Filter Coalescer control valve failing open dictates the maximum pressure at the spec break because it is located closer to the HP source.

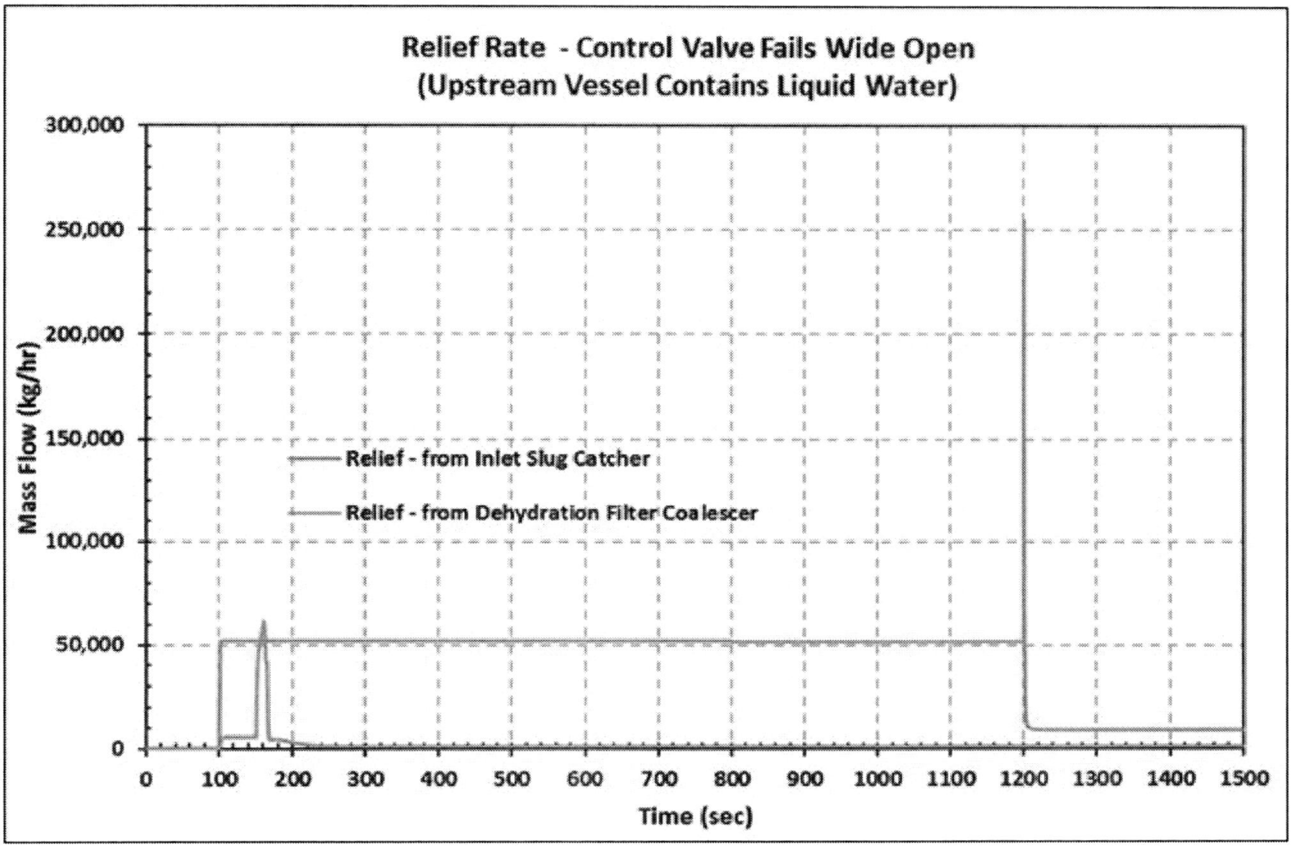

Figure 7. Relief Load from Slug Catcher and Dehydration Filter Coalescer

An existing facility identified that the pressure rating of piping may be insufficient in the event of a control valve failing wide open, resulting in vapor expansion. Dynamic simulation of all the circuits to the Stabilizer Feed Drum was performed to determine the maximum backpressure at the HP/LP interfaces. This evaluation allowed for the more cost-effective solution of retaining as much of the existing piping as possible. The results from the dynamic simulation were used to re-rate the piping and fittings and allowed for re-hydrotesting of segments rather than upgrading the entire system piping and adding pressure interlocks.

6.0 Conclusion

For a control valve failure scenario resulting in vapor expansion, the conventional methodology for determining the relief load does not consider any hydraulic losses in the system, resulting in a larger PSV size than may be required. The relief rate may be overestimated by considering liquid relief occurring at the volumetric rate of vapor expansion across the wide-open control valve without considering hydraulic losses.

Applying dynamic simulation to the same scenario, the relief load is limited based on the resistance in the system, resulting in a more accurate relief size requirement. Dynamic simulation is a valuable tool to also determine the maximum backpressure in the piping. For existing facilities, the results help to perform a fitness for service assessment.

513194 A CFD Study of System Vibration's Effect on LNG Boil-Off in a Transportation Vehicle

Abstract:

Liquefied Natural Gas (LNG) takes up 1/600 the volume of its gas form at normal temperature and pressure, which is suitable for long-distance transportation. However, as a cryogenic fluid, the very low temperature of LNG makes it very easy to evaporate and generate boil-off gas (BOG). For economic and environmental considerations, BOG should be minimized during transportations if possible. Lots of studies have been reported on the root causes and impacts of BOG. However, most of them focus on the factor of heat leakage into the cryogenic facilities to generate BOG. Actually, mechanical vibrations can also contribute to BOG generations, where quantitative studies are still lacking, especially on BOG generation effects during LNG vehicle transportations.

In this work, the vibration of LNG trucks has been considered to study BOG generations. The study with computational fluid dynamics (CFD) is performed to investigate the cylindric tank of an LNG truck, which was filled with LNG. The commercial CFD software, ANSYS Fluent V18, has been employed to develop the multiphase mixture model. Specifically, a user-defined function(UDF) has been programmed to model the vibration effect due to LNG truck transportation. The simulation has studied the effect of BOG generation under different vibration scenarios in terms of their frequency and magnitude. Based on the CFD simulation study, cost-effective BOG minimization methods during LNG vehicle transportations would be suggested.

Development of LNG cargo tank simulation technology for ballast voyage

Kenichi Matsuoka *Azbil corp.*
Carla Lara *Chemstations inc.*
Aaron Herrick *Chemstations inc.*
David Hill *Chemstations inc*

1. *Introduction*

In the Asia pacific region, Liquefied Natural Gas (LNG) is transferred by LNG vessels with cryogenic tanks intended for long-distance transportation. A significant fraction of the cargo LNG volume evaporates during the voyage as Boil-off Gas (BOG) particularly during spray operation. It is difficult to utilize generated BOG for LNG vessel operation efficiently, and especially to minimize (optimize) the LNG heel stock during a ballast voyage.

The authors developed a dynamic simulation system to estimate LNG heel stock in LNG cargo tanks of a Moss cargo containment system, in order to decide the required minimum LNG heel stock for a ballast voyage. The dynamic simulation is performed by a commercial process simulator (CHEMCAD). The simulation model considers the effect of several process factors such as LNG compositions, ambient temperature, pressure, sloshing factor (ship motion), spray operation schedule, and LNG vessel voyage plan. Based on the dynamic simulation, the optimum LNG heel stock can be decided for long-distance LNG transportation. The simulator uses simultaneous modular capabilities to solve the hydraulic balances between pressure and flow on the network during each time step of the dynamic simulation.

In this paper, we describe the rigorous and realistic dynamic simulation of boil-off gas, considering cargo tank spray operation and vessel energy requirements.

2. *Outline of heel stock estimation simulator for ballast voyage*

2.1 Purpose

As shown in Figure 2.2.1, each cargo tank has its own spray nozzle line to cool down its own gas phase temperature and metal wall temperature. Before LNG loading into the LNG vessel, the tank wall temperature at the tank equator is to be cooled down to at least -120 [degC], in order to avoid metal brittle fractures. It is the reason why the LNG tank spray operation is required during ballast voyage. In addition, fuel gas has to be supplied to the engine system (or boiler system) in order to satisfy the fuel energy demand required by the LNG vessel.

This process simulator can be utilized to estimate the required LNG heel stock for not only offline ballast voyage as planning tool, but also online ballast voyage as online spray control tool.

2.2 Scope of process

Figure 2.1.1 shows the scope of processes to be conducted by dynamic process simulation. The most important operation during the ballast voyage is to cool down the LNG cargo tank wall (Moss) at the end of the ballast voyage while satisfying the required fuel demand during the entire voyage.

The process flow model is composed of the following equipment:

(1) LNG cargo tank

(2) LNG forcing vaporizer (F/V)

(3) Spray pump

(4) Piping system

(5) Control system

(6) Fuel demand system (to engine or boiler for propulsion system)

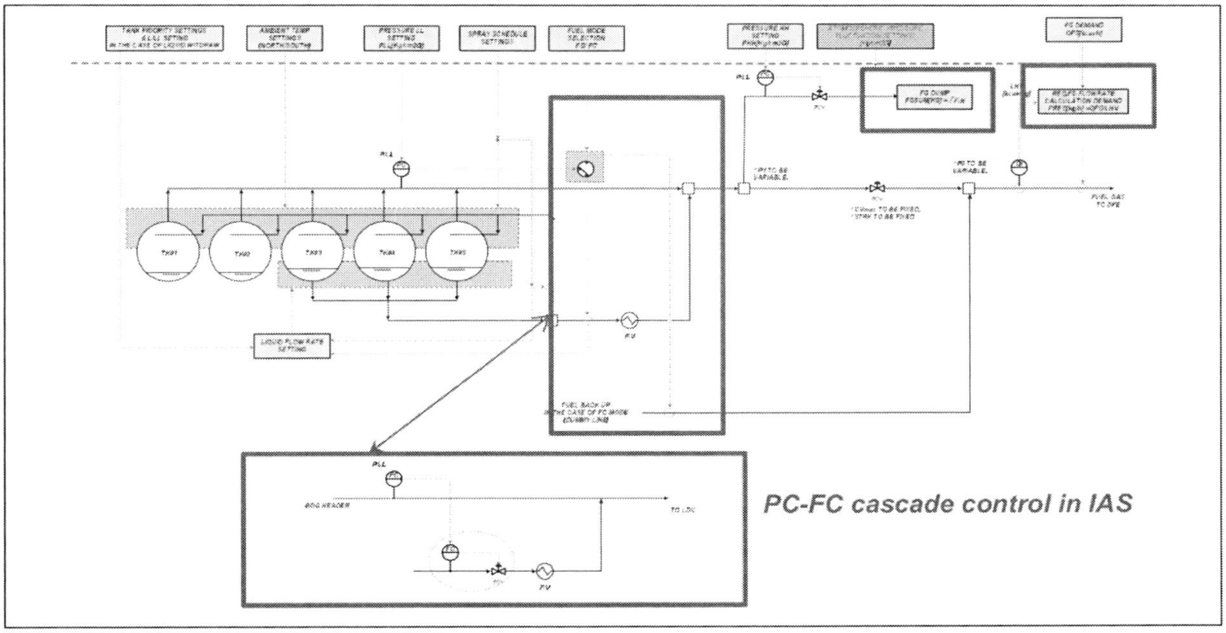

Fig.-2.2.1: Dynamic process modelling schematic for Cargo tank system

2.3 Process model

Herein, the concept of cargo tank model and the related control logics are shown in

Fig.2.3.1. The LNG cargo tank model is composed of the following sub-process models:

A) *Heat input from LNG cargo tank wall*

B) *Heat input by spray pump*

C) *Heat accumulation on LNG cargo tank wall*

D) *Heat accumulation on LNG liquid phase*

E) *BOG generation from the interface between liquid and vapor*

F) *Heat accumulation on LNG vapor phase*

G) *LNG mist M/B & H/B in tank vapor phase by spray nozzle*

H) *Spray efficiency*

(1) *Cargo tank model:*

Fig.2.3.1 shows the simplified tank model system of BOG generation from LNG cargo tank.

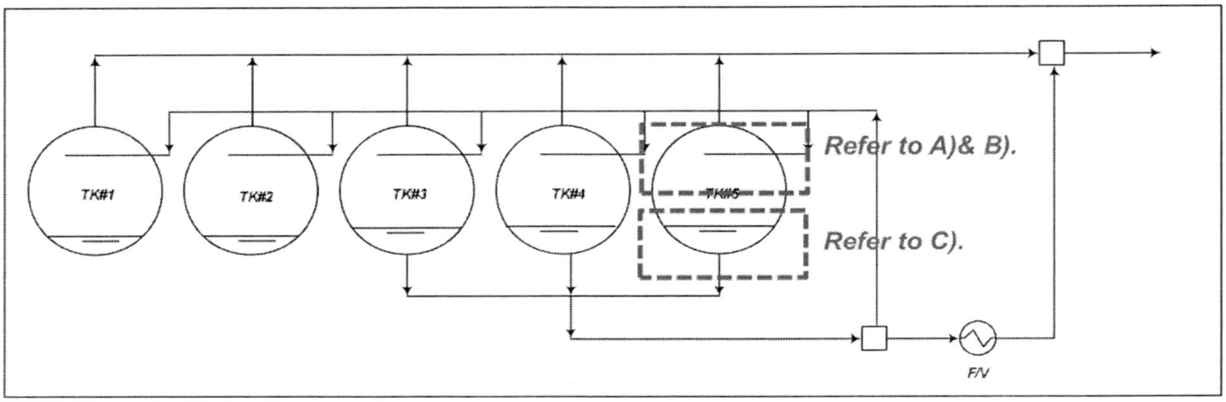

Fig.-2.3.1: Simplified cargo tank system

A) *Vapor phase model:*

Fig 2.3.2 shows the process behavior diagram of vapor phase in cargo tank.

Gas phase model and Mist evaporation model shown in the figure is described as follows:

Gas phase:

- *Gas Mass Balance:* $\therefore \frac{d}{dt}W_g(t) = F_{BOG}(t) + F_{vms}(t) - F_o(t)$

- *Mist Mass Balance:* $\therefore \frac{d}{dt}W_{ms}(t) = F_m(t) - F_{vms}(t) - F_{s,mist}(t)$

- *Energy Balance* : $\therefore \frac{d}{dt}\{H_g(T_g) \times W_g(t)\} = H_g(T_l) \times F_{BOG}(t) + H_l(T_l)$

$\times F_{vms}(t) - H_g(T_g) \times F_o(t) + Q_{in,g}(t)$

Mist evaporation model:

- *Mist flow rate:* $\therefore F_{w,mist} = (F_{pump} \times \rho_l) \times \eta_{spray}$
- *Mist evaporation rate:* $\therefore F_{w,vms} = Q_{w,vms} / (h_{w,vms} - H_{w,vms})$
- *Mist heat transfer rate:* $\therefore Q_{w,vms} = h_{mist} \times A_{mist,total} \times (T_g - T_l^*)$

Spray efficiency:

- *Efficiency:* $\therefore \eta_{spray} = f(\Delta T(\equiv T_g - T_l^*),\ \rho_l)$

Where, $W_g(t)$; *Gas weight in tank [kg]*

$F_{BOG}(t)$; *BOG rate from leaving from liquid phase [kg/h]*

$F_{vms}(t)$; *Vaporized gas from mist in tank [kg/h]*

$F_o(t)$; *Suction pure gas flow at tank (w/o mist flow) [kg/h]*

$W_{ms}(t)$; *Total mist weight in tank [kg]*

$F_m(t)$; *Mist flow rate by spray [kg/h]*

$F_{vms}(t)$; *Vaporized gas from mist in tank [kg/h]*

$F_{s,mist}(t)$; *Entrained mist flow rate in suction gas flow [kg/h]*

$H_g(T_g)$; *Gas enthalpy at Tg [degC] (vapor phase temperature)*

η_{spray} ; *Spray pump efficiency [-]*

T_g ; *Gas phase temperature [degC]*

T_l^* ; *Mist temperature (=LNG spray temperature) [degC]*

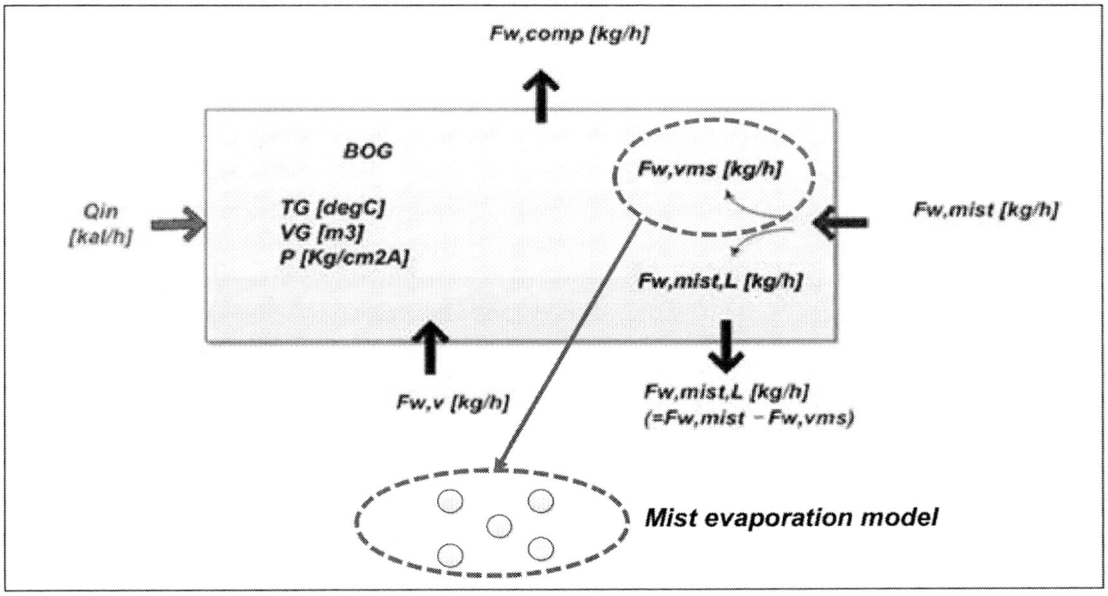

Fig.-2.3.2: Vapor phase model

B) Tank wall heat balance model:

Fig.2.3.3 shows the process behavior diagram of tank wall in cargo tank.

Heat transfer between ambient and vapor phase (trough tank wall) are shown as follows:

- *Heat balance:*

$$\therefore C_{pw}W_w\frac{dT_w}{dt} = h_{o,ins}A_o(T_{amb} - T_w) - h_iA_i(T_w - T_g)$$

- *Heat transfer coefficient:*

$$\therefore 1/h_{oi,ins} = 1/h_o + 1/(k_{ins}/t_{ins})$$

Where, C_{pw} ; *Gas weight in tank [kg]*

W_w ; *BOG rate from leaving from liquid phase [kg/h]*

$h_{o,ins}$; *Effective heat transfer coefficient between ambient and wall Including insulation [kcal/(m2.h.degC)]*

h_i ; *Heat transfer coefficient between wall and gas phase [kcal/(m2.h.degC)]*

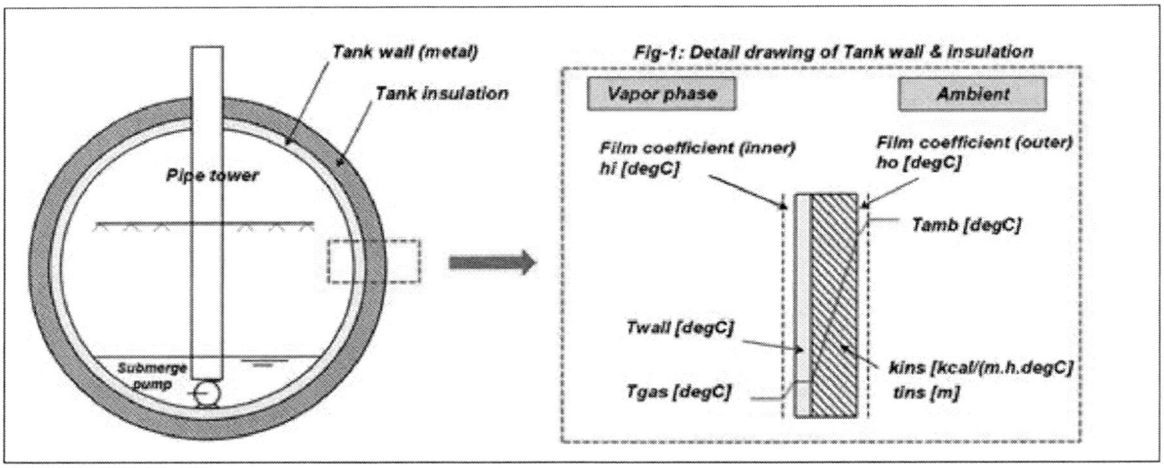

Fig.-2.3.3: Tank wall model

C) Liquid phase model in tank

Fig.2.3.4 shows the process behavior diagram of liquid phase in cargo tank.

Liquid phase model shown in the figure is described as follows:

Liquid phase:

- *Mass Balance:* $\therefore \rho_l\frac{d}{dt}V_l(t) = -\{F_{BOG}(t) + F_m(t)\}$

- *Energy Balance:*

$$\therefore \frac{d}{dt}\{H_l(T_l) \times W_l(t)\} = -\{H_g(T_l) \times F_{BOG}(t) + H_l(T_l) \times F_m(t)\} + Q_{in,l}(t)$$

Where, $V_l(t)$; LNG volume in the tank [m3]

ρ_l ; LNG density [kg/m3]

$F_{BOG}(t)$; BOG rate leaving from liquid phase [kg/h]

$F_m(t)$; Mist flow rate leaving from spray nozzle [kg/h]

$H_l(T_l)$; Liquid enthalpy at Tl [degC] (liquid phase temperature)

$H_g(T_l)$; Gas enthalpy at Tl [degC] (liquid phase temperature)

$Q_{in,l}(t)$; Heat input to tank liquid phase [kcal/h]

T_l ; Liquid phase temperature [degC]

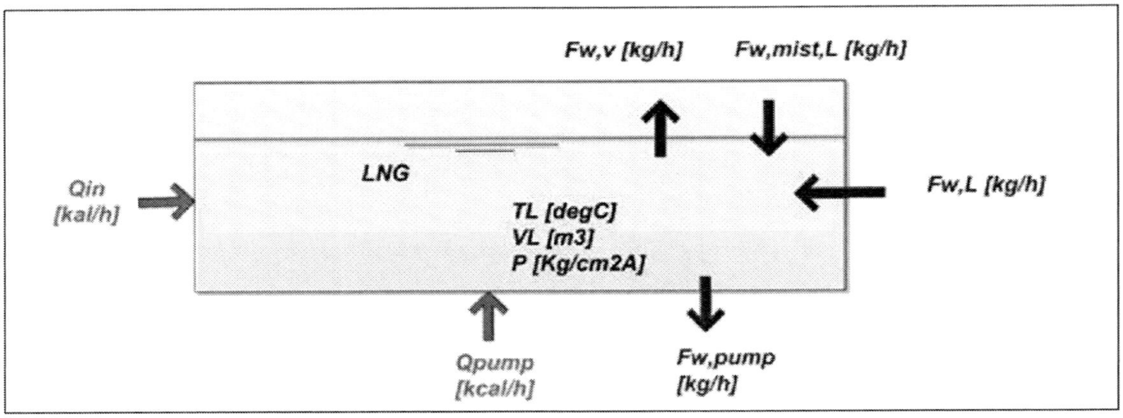

Fig.-2.3.4: Liquid phase model

(2) Control logics:

The LNG vessel model involves not only BOG & spray mist calculation to find tank wall and gas phase temperatures, but also the following control sequences:

A) Tank pressure HH/LL logic:

- Pressure HH:

  Excess BOG is to be dumped, in order to protect LNG cargo tank.

- Pressure LL:

  F/V is to start during FG supply mode, and vaporized natural gas is to supplied to engine (or boiler)

B) Tank LNG level LL logic:

- If spray pump related the tank with level LL is running, the pump is to stop in order to protect itself. Then, sequentially a pump in another tank with enough LNG inventory starts.

C) FO & FG mode operation:

- FO or FG priority combustion mode is to be selected. Iin the case of FO selection mode, F/V is not utilized.

2.4 System configuration

The integrated system configuration for the dynamic process simulator is shown as follows:

A) Dynamic simulation (process model, control model)

B) Simulator GUI (GUI, LNG vessel data, simulated database)

Calculation result (at every time step) is sent to the upper system (through I/F), in order to show the dynamic behavior.

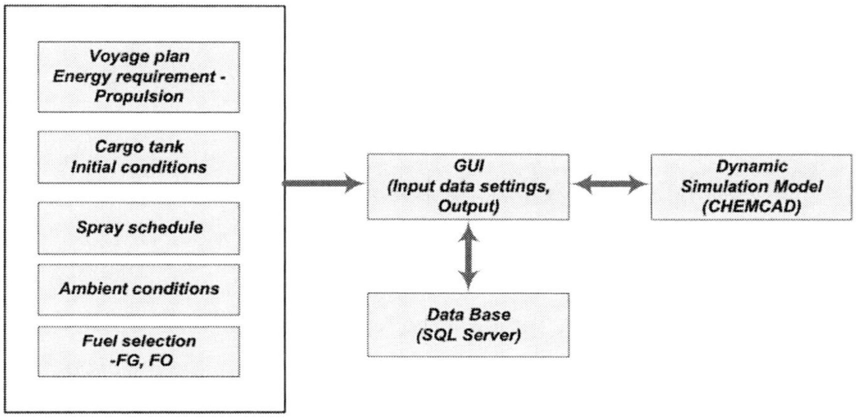

Fig.-2.4.1: Configuration of Dynamic Simulation System

3. *Dynamic process simulation on heel stock estimation*

3.1 Case study

The simulation scenario to be executed here is taken as follows:

(1) Step-1: Validation of process simulation model

The dynamic process model shown in Figure 2.1.1 was validated based on the actual ballast voyage data including spray operation.

Herein, the following two ballast voyage simulations, which have each different LNG compositions as heel stocks, were executed in order to show whether the dynamic simulation model is able to represent the actual process behavior.

A) Qalhat LNG: Ballast voyage from Japan to Qatar

B) Barrow LNG : Ballast voyage from Japan to Australia

The results of the above dynamic simulations are shown in section 3.2.

Table-3.1.1: LNG composition data

			A) Ballast voyage	B) Ballast voyage
			Qalhat LNG	Barrow LNG
1	Methane	[mole%]	90.08	94.59
2	Ethane	[mole%]	6.01	3.85
3	Propane	[mole%]	2.27	0.92
4	Iso-butane	[mole%]	0.77	0.19
5	N-Butane	[mole%]	0.69	0.21
6	Iso-Pentane	[mole%]	0.01	0.03
7	N-Pentane	[mole%]	0.00	0.00
8	Hexane	[mole%]	0.00	0.00
9	Nitrogen	[mole%]	0.17	0.21
10	CO2	[mole%]	0.00	0.00
	Total	[mole%]	100.00	100.00

(2) Step-2: Effect of spray schedule on tank process behaviors

As explained in section 2.1, during ballast voyage, the metal wall temperature of the LNG cargo tank is cooled down to at least -120 [degC] before loading LNG into the vessel at LNG plant by LNG cargo tank spray operation.

In order to find the optimum spray schedule to consume the lowest amount of spray flow, the sensitivity study, which is the effect of spray schedule on tank process behavior such as the change of LNG inventory, metal wall temperature etc., was executed.

Herein, based on the simulation result of the above "A) Qalhac LNG ballast voyage", the voyage simulations for different spray schedules were executed, in order to show the effect of spray schedule on cargo tank process behavior.

The result of simulations are sown in section 3.3

3.2 Process Study-1: Dynamic model validation

The ballast voyage simulations based on the following LNG kind and voyage duration were conducted, in order to validate the accuracy of dynamic process model. The results of process behavior are shown as follows:

- *A) Qahtac LNG ballast voyage :Duration = 350 [h] (=14 days)*
- *B) Barrow LNG ballast voyage: Duration = xxx*

3.21 Effect of change of LNG on spray efficiency

Before executing the dynamic process simulation described in 3.2.2, the spray efficiency for each LNG composition data shown in table 3.1.1 were analyzed based on the actual process data.

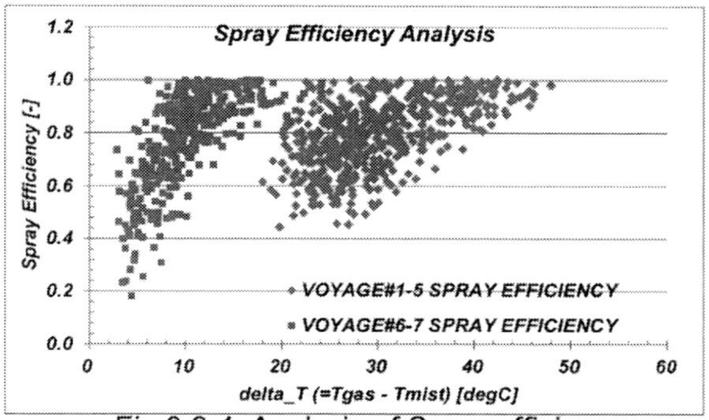

<Notes>:

(1) Voyage#1-5: Qalhat LNG

(2) Voyage#6-7: Barrow LNG

Fig.3.2.4: Analysis of Spray efficiency

It is observed that there are different characteristics of spray efficiency due to the difference of LNG composition,

3.2.2 Process behavior in LNG cargo tank

Based on the analysis data of spray efficiency shown in 3.2.1, the ballast simulations for each LNG type (A)Qalhat LNG, B)Barrow LNG) were executed to validate the accuracy of the dynamic process model. The calculation results and the related actual voyage data are shown as follows:

(1) Tank wall temperature:

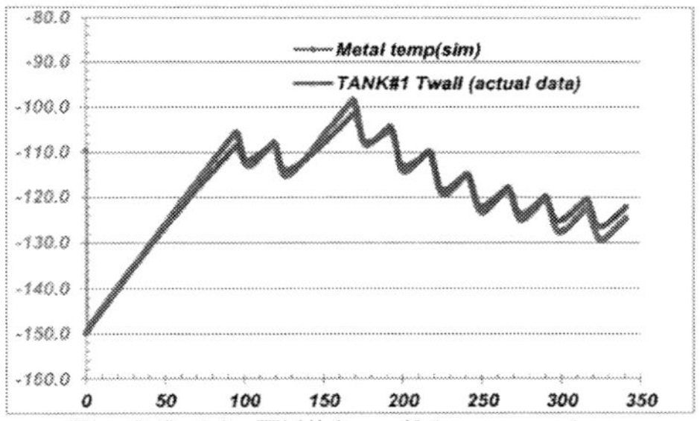

Fig.3.2.1A: TK#1 wall temperature

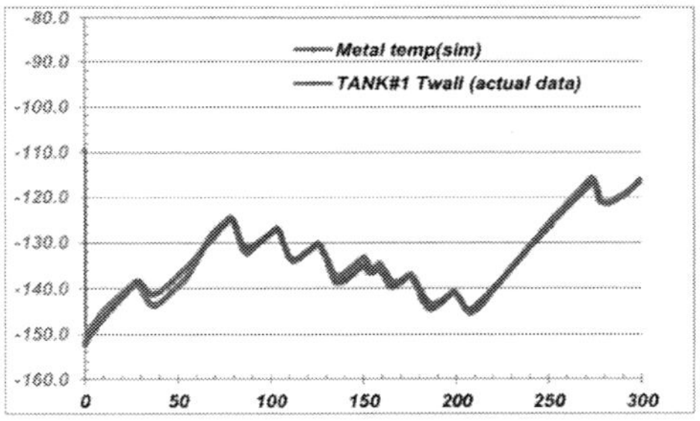

Fig.3.2.1B: TK#1 wall temperature

(2) Gas phase temperature:

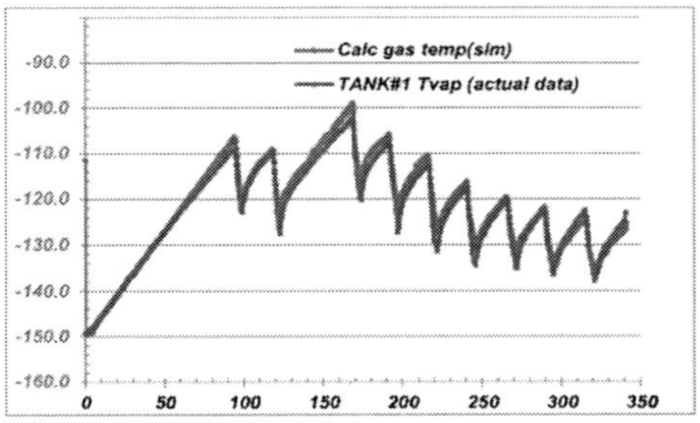

Fig.3.2.2A: TK#1 gas temperature

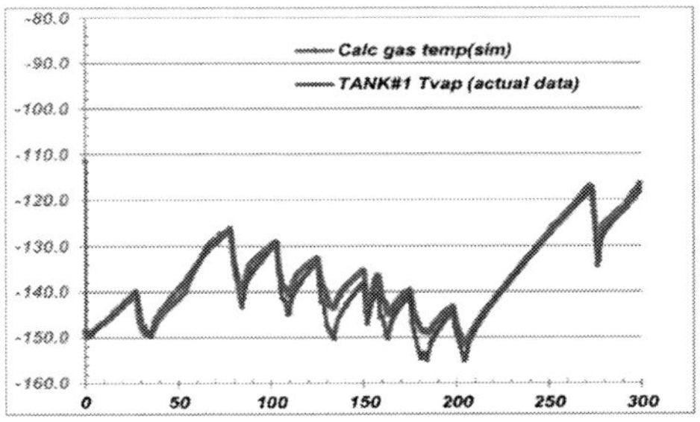

Fig.3.2.2B: TK#1 gas temperature

(3) LNG inventory:

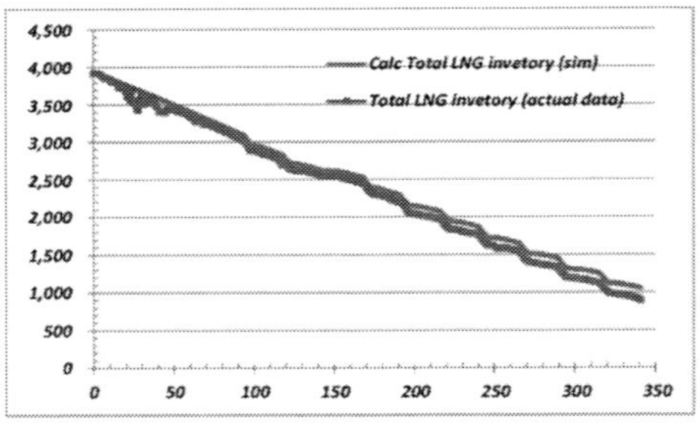

Fig.3.2.3A: TK#1-5 Total inventory

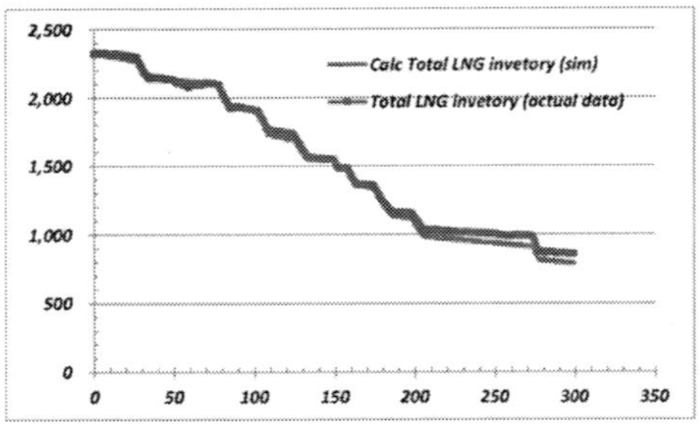

Fig.3.2.3B: TK#1-5 LNG inventory

3.3 Process Study-2: Effect of spray schedule on tank process behavior

In order to analyze the effect of change of spray schedule on cargo tank process behavior, ballast voyage simulations were executed.

Herein, in order to focus on the estimation of the net BOG generation rate and the required spray flow rate to satisfy the required minimum tank wall temperature at the end of the ballast voyage, it is assumed that FG demand flow rate for this study is zero.

The dynamic simulations based on the following spray schedule were conducted:

A) *Base Case: Actual spray schedule*
B) *Case-1: Spray operation conducted throughout the voyage*
C) *Case-2: Spray operation conducted for the final part of the voyage*

(1) Tank wall temperature & BOG rate from tanks:

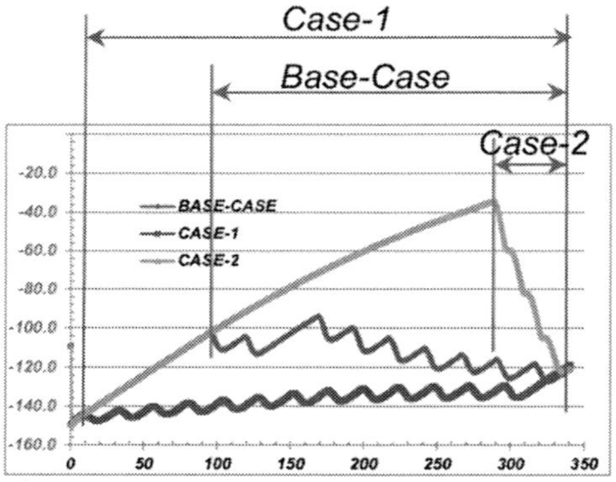

Fig.3.3.1: TK#1 wall temperature Due to change of spray schedule

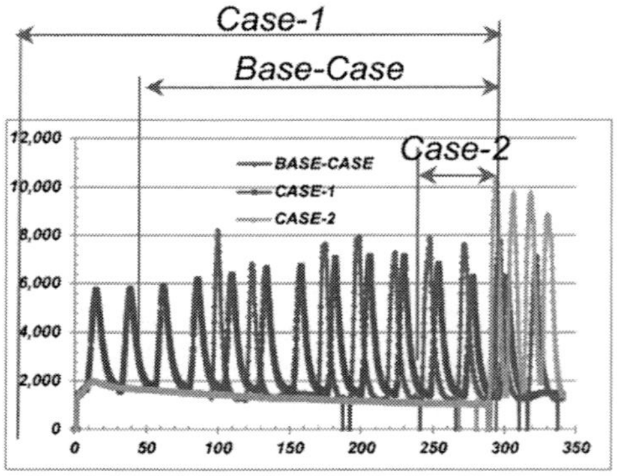

Fig.3.3.2: BOG generation rate from cargo tanks Due to change of spray schedule

(2) Total LNG inventory in tanks & Integration of BOG flow rate:

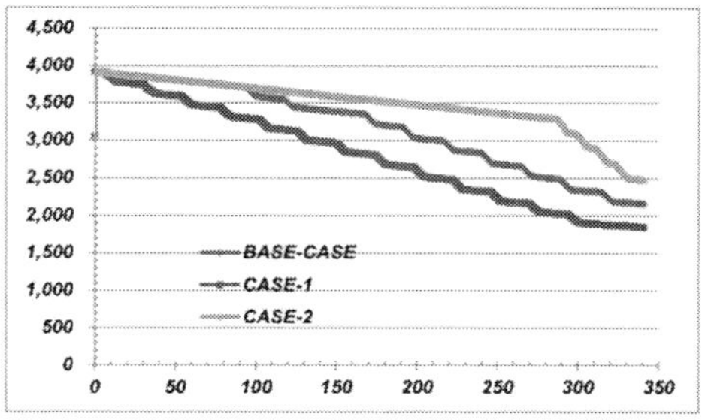

Fig.3.3.3: Total LNG inventory in cargo tanks Due to change of spray schedule

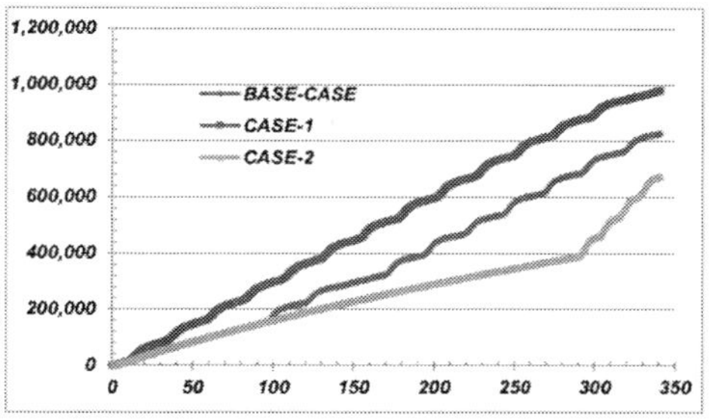

Fig.3.3.4: Integration of BOG generation rate Due to change of spray schedule

Based on the above calculation results, BOG generation rate for each spray schedule is summarized as follows:

Table.3.3.4: Integration of BOG generation rate

			BASE-CASE	CASE-1	CASE-2
TOTAL BOG GENERATION RATE	W	[kg]	826,757	979,245	673,831
TOTAL BOG GENERATION RATE	V	[m3]	1,805	2,138	1,471
	Δ V	[m3]	0	333	-334
TOTAL BOG GENERATION RATE		[%]	100.0	118.4	81.5
(*1) LNG density =458 [kg/m3]					

As shown in Table 3.3.4, CASE-1 scenario, which is the longest duration of spray schedule to keep the cargo tank at lower temperature during the entire ballast voyage, causes maximum BOG generation among three cases, and CASE-2 scenario, which is the centralized spray operation at the end of ballast voyage, causes minimum total BOG generation rate. In terms of heat transfer rate into tanks, cargo tanks with lower vapor temperature have higher heat intake through tank wall, which leads to higher required spray flow amount. The optimum spray schedule with minimum amount of spray flow to cool down tank wall should be decided by taking into consideration optimum voyage route and keeping higher propulsion performance.

4. Conclusion

4.1 Summary

A dynamic process simulator with spray model for LNG cargo tank system during ballast voyage has been developed, in order to estimate the required LNG heel stock and make a plan for optimum spray operation schedule during ballast voyage

The dynamic model can be applied for:

(1) LNG ballast handling (including spray schedule) during voyage

(2) Thermo-economical voyage simulation of marine energy systems of LNG

vessels

The LNG cargo tank model is a dynamic process model for the evaporation of LNG during marine transportation. The main benefit of this model is that it is able to simulate boil-off gas rate and the quantity of LNG stock in the tanks, and also execute the dynamic process simulation for the energy system of LNG vessels.

4.2 Way forward

We are focusing on the development of an online dynamic simulation system (on-board usage), in order to find the optimum spray schedule and predict the change of LNG heel stock in cargo tanks. The concept of the system is as follows:

(1) Real-time process parameter estimation to compensate dynamic process model:

- *Heat transfer coefficient (gas phase, liquid phase)*
- *Spray flow rate*
- *Spray efficiency*

The process parameters are estimated based on the following optimization calculations, and reflected into the dynamic simulation model.:

$$\therefore Minimize \left|V_{liq,cal}(t) - V_{liq,act}(t)\right| \rightarrow Estimate\ U_{liq}[kcal/(m2.h.degC)]$$

$$\therefore Minimize \left|T_{wall,cal}(t) - T_{wall,act}(t)\right| \rightarrow Estimate\ U_{gas}[kcal/(m2.h.degC)]$$

(2) Real-time dynamic simulation to predict process behavior in cargo tanks by integration of voyage planner and LNG tank simulator:

- *Cargo tank equator temperature (metal temperature)*
- *BOG generation rate*
- *LNG Inventory*

This dynamic simulator, which is able to predict the behavior of LNG cargo tank process variables shown above, will be a part of a Ship Information Management System (SIMS) with the aim of precisely understanding the fuel consumption and propulsion performance conditions of oceangoing vessels.

As shown in Fig 4.2.1, it will be combined with Voyage planner (to find the optimum voyage route and vessel speed) on LNG vessels to save more fuel energy. The Gas Management System (GMS) on LNG vessels will be able to be operated more effectively by utilizing the integrated dynamic simulator (as online system), based on the actual vessel operation data (vessel speed and fuel consumption, predicted weather conditions)

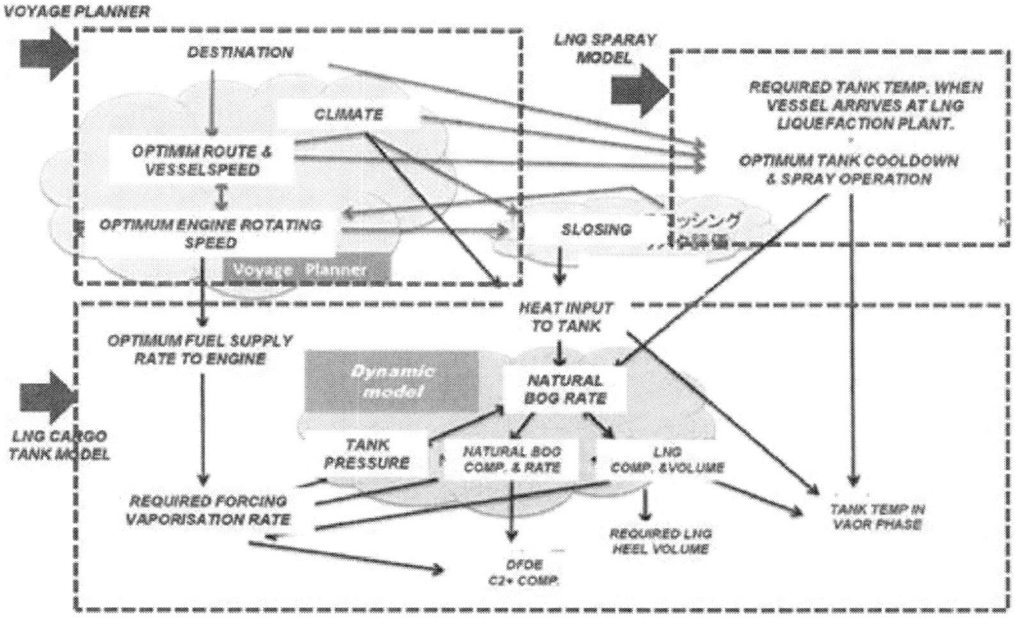

Fig.4.2.1: Integration of voyage planner and LNG cargo tank simulator (Adapted from the concept of voyage planner system by MTI Co.Ltd)

< Acknowledgments>

The work is supported by Nippon Yusen Kabushiki Kaisha (NYK) and MTI Co. Ltd participating project, selected by the Ministry of Land, Infrastructure, Transport and Tourism (MLIT) for "2016 Support Projects for R&D in Advanced Safety Technology of Vessels."

<Bibliography>

1. *N.Umeyama, M.Kawashima, Y.Nakamura, N.Umeyama, "BOG CONTROL SIMULATION DURING BALLAST VOYAGE", International Conference on Liquefied Natural Gas ;C-1, in 1995*
2. *N.Umeyama, K.Matsuoka, A.Iwata, D. Hill, "Development of dynamic process simulator on LNG vessel gas management system", AICHE 15th Topical Conference on Gas Utilization, in 2015*
3. *Luyben, "Process modeling, simulation and control for Chemical engineer", second edition, McGraw Hill, USA*

176d

High Capacity Concept for an Efficient Gas Expander FLNG Process

Heinz Bauer (speaker), Linde Engineering, Pullach, Germany
Frank Hänsel, Cryostar, Hésingue, France
Stefano de Simone, BHGE, Florence, Italy

Abstract

LNG plants with small-to-medium capacity rely frequently on refrigeration via work expansion of light gases. In many cases nitrogen is used as working fluid, but methane is also considered as eligible candidate. To achieve a fair efficiency gas expander processes are designed in such a way that two or more expanders are operated in different temperature ranges. Thus, the cooling duty for sub-cooling, liquefaction and in some cases, even pre-cooling of natural gas is provided by dedicated machines.

In a widely-used concept, which had been introduced first by Chris Dubar (e.g. expired US patent 5,768,912) two nitrogen expanders operate over the same pressure profile, but generate refrigeration at different temperatures, as the inlet temperature into the expanders is adjusted accordingly. A good thermodynamic process efficiency requires a narrow approach of heating and cooling curves. As the cooling duties are different in the respective cooling steps the expander operated at the cold end of the process will contribute only with a relatively small share to the overall refrigeration duty.

Obviously, the capacity of a single LNG train is governed by the maximum size of each piece of equipment. In a traditional dual gas expander process this bottleneck is established by the warm expander, which must provide 70-85% of the overall duty. An enforced duty balance between warm and cold nitrogen expander would cause an unacceptably high drop in process efficiency.

The objective of a joint development was to balance the duty of the warm and cold expander without sacrificing thermodynamic efficiency. As this goal cannot be reached with pure nitrogen as working fluid the well-known methane/nitrogen mixture was investigated. This paper will discuss in detail the features of the patented process (US patent 9,841,229), which is marketed under the name LIMEN™ (LInde MEthane Nitrogen process).

When two large expanders (e.g. Cryostar size TC600/130) and a large BHGE aeroderivative gas turbine LM9000 are integrated into the LIMEN™ process, a double train LNG capacity of more than 2.5 mtpa LNG can then be achieved. This concept is a major step-forward for floating LNG projects, where economy-of-scale and lowest possible CAPEX are of the essence.

Introduction

Floating natural gas liquefaction (FLNG) projects are still exotic amongst more than hundred mid and world scale onshore LNG trains with at least 1 mtpa capacity. Only the Petronas floater #1 (PFLNG Satu)[1] and SNH/Perenco's Kribi (Hilli Episeyo) project[2] are in operation yet with Shell's flagship 'Prelude'[3] ready for start-up in 2018. With reported specific CAPEX figures[4] between 2,000 and 3,000 US$ per annual ton of LNG for realistic FLNG projects there is still room for cost reductions to get competitive with e.g. shale gas based LNG projects. A KPMG report[5] provides further insight into crucial aspects of successful FLNG projects.

CAPEX reductions can be achieved by increasing the liquefaction capacity to improve the economy of scale. Not every time proven onshore concepts translate into optimal FLNG solutions. In many cases excellent operability is more important than outstanding efficiency. This offers new opportunities for expander based liquefaction processes.

State-of-the-art N2 expander processes

Generic process

About 20 years ago BHP announced during LNG12 in Perth a new concept for world scale LNG production[6]. The first application of the cLNG process could have been a 2-3 mtpa gravity based facility, which should have produced gas from the Bayu-Undan field. BHP, however, was not successful to convince the other shareholders about the benefits of the new approach. BHP sold its interests to (Conoco)Phillips, which is producing nowadays the Bayu-Undan gas in Darwin/Australia.

What was new with the cLNG process (Figure 1)? Based on well-proven work expansion of nitrogen in the vapor phase only, the combination of two expanders operating in different temperature ranges reduced the energy consumption of the process to an acceptable level. In comparison to more efficient phase-change processes based on pure or mixed refrigerants the cLNG process does not require suction drums for compressors or pumps. Further, nitrogen is readily available also in remote locations and does not raise safety concerns.

[1] http://www.petronas.com.my/media-relations/media-releases/Pages/article/PETRONAS%E2%80%99-FIRST-FLOATING-LNG-FACILITY,-PFLNG-SATU.aspx

[2] https://www.offshoreenergytoday.com/perenco-starts-up-hilli-episeyo-flng-off-cameroon/

[3] https://www.shell.com/media/news-and-media-releases/2017/prelude-flng-begins-its-journey-to-australia.html

[4] Energy World Shipping, Gaffney, Cline & Associates; Excelerate Energy and Oxford Institute for Energy Studies, Infographic by Richard Neighbour

[5] https://assets.kpmg.com/content/dam/kpmg/pdf/2014/11/floating-LNG-evolution-and-revolution-for-the-global-industry.pdf

[6] http://www.ogj.com/articles/print/volume-96/issue-23/in-this-issue/gas-processing/bhp-unveils-new-compact-lng-technology.html

A tight heat integration between warm and cold streams of the cLNG process requires a large duty of the warm expander, which acts as pre-cooling and liquefaction refrigerant provider, and a small duty of the cold expander, which serves as sub-cooling refrigerant generator. Depending on actual conditions like feed gas pressure and composition the duty of the warm expander may be as high as 70-85% of the overall duty causing a significant imbalance between the warm and the cold expander.

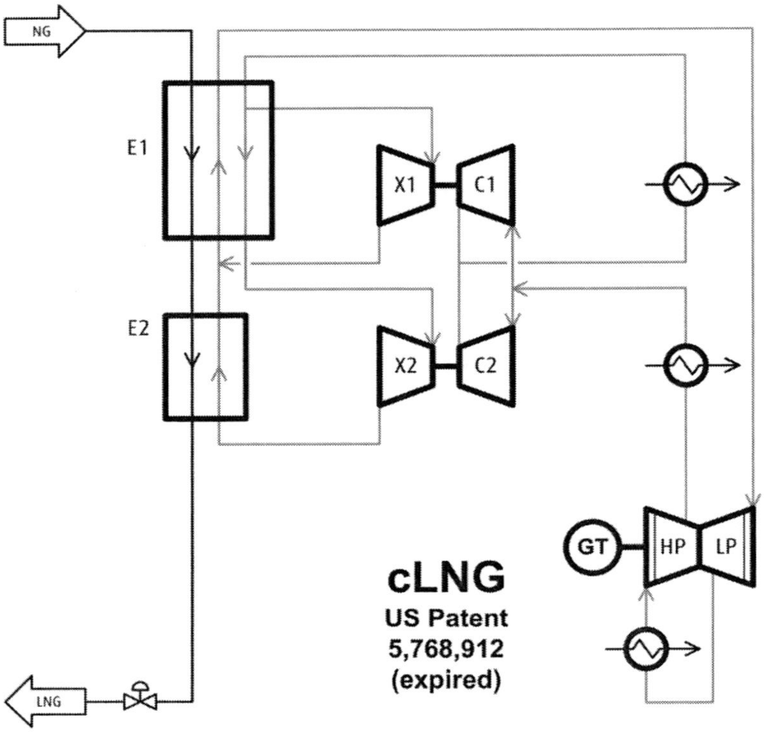

Figure 1 double N2 expander process (cLNG)

If the warm expander does not hit limits with respect to size and duty, the asymmetry between both expanders is not a big concern. LNG plants using this (meanwhile public domain) concept are widely used in the industry, mostly for small scale applications like the LNG facility[7] for GazMétro in Montreal/Canda, which has been built and started-up in 2017 by Linde[8] as EPC contractor.

[7] https://canada.constructconnect.com/dcn/news/projects/2017/05/gaz-metro-wraps-up-120-million-lng-plant-expansion-in-montreal-1023573w

[8] https://www.linde-engineering.com/internet.global.lindeengineering.global/en/images/Star_LNG19_166489.pdf?v=5.0

Petronas FLNG process

It took more than 20 years between filing the patent of the cLNG process and first LNG[9] production of the Petronas floater #1 (PFLNG Satu). PFLNG Satu, however, does not use the process as originally invented by BHP, but applies a modified APCI design[10] (Figure 2). Here up to three expanders are operating with two different ratios between inlet and outlet pressure. Even if a detailed comparison between cLNG and AP-N is not publicly accessible an efficiency improvement from cLNG to AP-N can be expected. A common feature, however, between cLNG and AP-N is the use of expanders with significantly different shaft power.

The two Petronas floaters Satu and Dua are not twins, rather sisters. PFLNG Satu uses GE PGT25+G4 gas turbines to produce 0.6 mtpa per train (1.2 mtpa in total) while PFLNG Dua applies GE LM6000PF as mechanical drives for a liquefaction process with 0.75 mtpa per train (1.5 mtpa in total)[11].

So far, all FLNG projects are designed with at least two independent liquefaction trains. This concept provides more flexibility with respect to start-up and production quantities during scheduled and unscheduled plant outages.

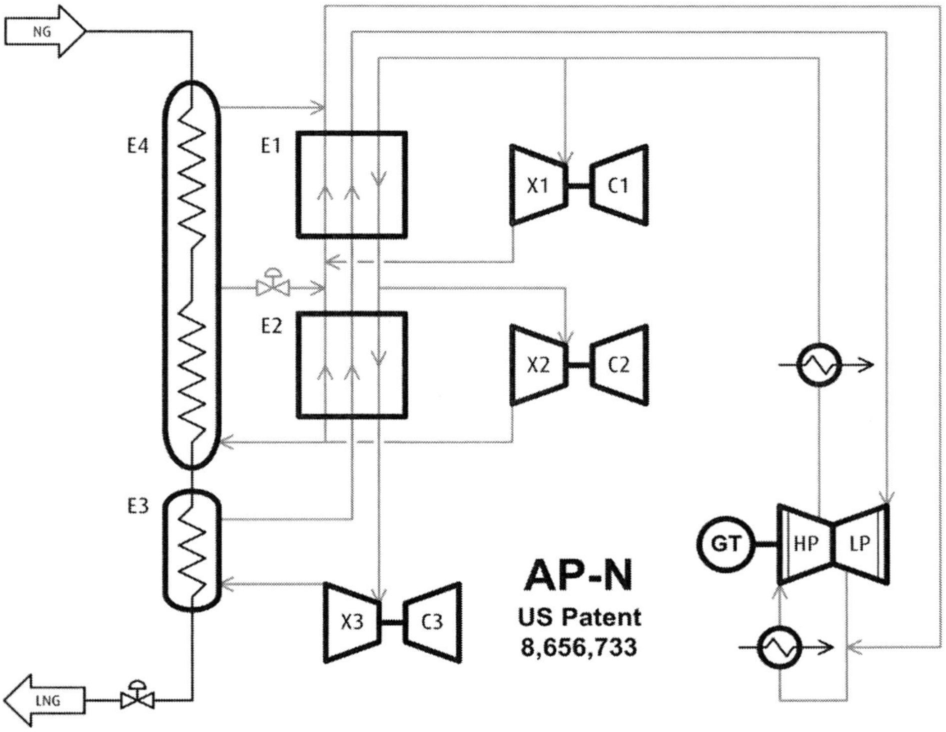

Figure 2 enhanced N2 expander process (AP-N)

[9] http://www.petronas.com.my/our-business/Upstream/lng/Pages/PFLNG.aspx
[10] http://www.airproducts.com/industries/energy/lng/~/media/0DEF92F964E04F2EAE5D39A9CD1AFFF2.pdf
[11] http://www.genewsroom.com/press-releases/ge-oil-gas-supplying-malaysia%E2%80%99s-petronas-turbomachinery-second-floating-lng-facilit-0

Methane/nitrogen expander processes

As explained above nitrogen expander processes inherently require different expander sizes, if an acceptable thermodynamic efficiency is expected. Once the train capacity must be maximized new approaches need to be taken to overcome prevailing design limits. Two dogmas of traditional nitrogen expander processes must be sacrificed to offer more headroom for design improvements

- Replace pure nitrogen by a methane containing mixture
- Accept condensation and vaporization (phase change) of the refrigerant

Even though other components might be conceivable the use of methane as partner of nitrogen is straight forward for an LNG plant with ample availability of high purity methane. Several process configurations with a CH4-N2 mixture have been discussed by Mokhatab et al[12]. Condensation of a portion of the HP refrigerant and re-vaporization at the cold end of the process allows to shift the operation of the cold expander to a warmer temperature range.

Single expander process

The use of a CH4-N2 mixture as refrigerant has been very popular for peak shaving LNG plants, where the ease of start-up and operation based on one single expander only was very much appreciated. One example of the technology is shown in Figure 3.

Figure 3 LNG peak shaving plant Stuttgart/Germany

[12] Saeid Mokhatab, John Y. Mak, Jaleel V. Valappil, David A. Wood, Handbook of Liquefied Natural Gas, Elsevier, 2014, page 220 ff

LIMEN™ (Linde Methane/Nitrogen) process

The patented (US patent 9,841,229) LIMEN™ process relies on proven building blocks, which are compiled in a novel and sophisticated manner. The simple and clean process layout quotes features from the cLNG process. The use of a methane/nitrogen mixture as refrigerant and a phase-change step in E3 (Figure 4) shows the following effects

- In E3 sub-cooling is provided by efficient condensation/vaporization like the sub-cooler design of well-known mixed refrigerant processes
- The operating range of the cold expander X2 is shifted from sub-cooling service to a warmer range where more refrigeration can be used efficiently
- With proper selection of the operating conditions of the feed gas similar or even identical shaft power of X1 and X2 can be achieved at the thermodynamic optimum (=minimum energy consumption) of the process

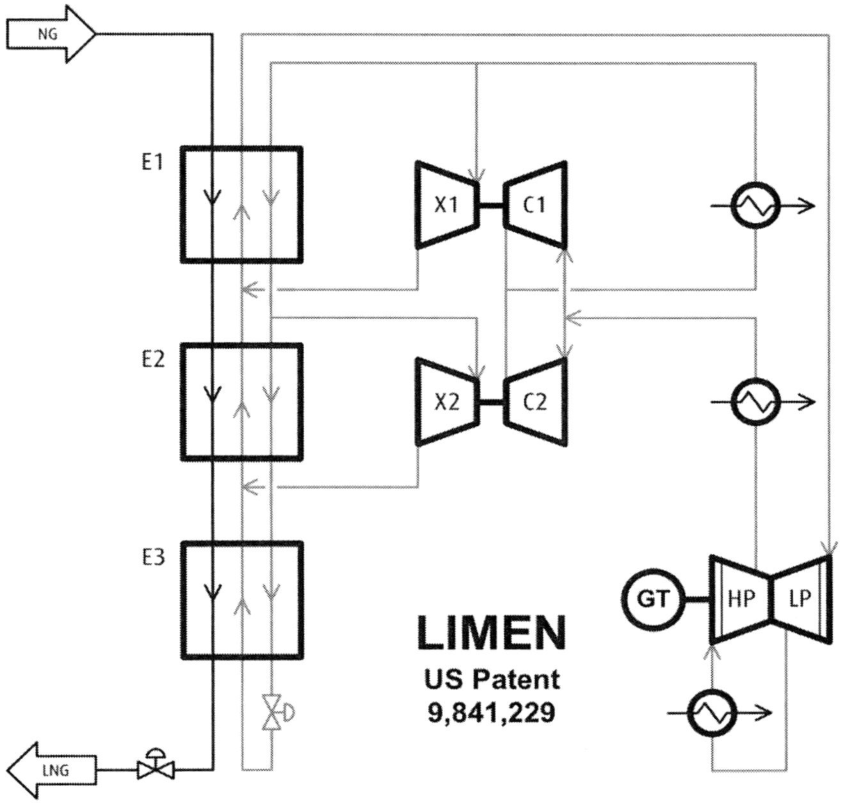

Figure 4 symmetric CH_4/N_2 expander process (LIMEN™)

The LIMEN™ process does not contain liquid hydrocarbons except a minor volume at the cold end of E3, which is equivalent to the inventory of HP LNG upstream of the JT valve. The narrow range of flammable CH_4/N_2 mixtures and the absence of a BLEVE[13] risk explains the excellent safety features, which are on the same level as a nitrogen expander process.

[13] https://en.wikipedia.org/wiki/Boiling_liquid_expanding_vapor_explosion

Rotating Equipment for the LIMEN™ process

Smart process concepts are of little value, if suitable equipment is not available. While static equipment (mostly brazed aluminum heat exchangers and inter-/aftercoolers) are readily available the selection of appropriate rotating equipment is more challenging.

Capacity per train (mtpa)	Shaft power per expander (MW)	Expander model (Cryostar)	Gas turbine driver (BHGE)	GT power @ 7°C (MW)	Recycle compressor (BHGE)
0.60	7.4	TC400/90	LM2500+G4 or PGT25+G4	>31	2BCL1006
0.95	11.6	TC500/110	LM6000PF+	52	2BCL1206
1.35	16.6	TC600/130	LM9000	70	2BCL1406

Table 1 selected rotating equipment for various plant sizes

Turboexpander/Compressor (Cryostar)

The industrial application of turboexpanders (also called 'expander') has been common since 1939 for cryogenic processes where expansion of a gas performing work results in a reduction in enthalpy of the working medium[14]. An expander is a machine that continuously expands a medium from a high pressure to a lower pressure resulting in an enthalpy drop. This drop in enthalpy corresponds to the increase in mechanical energy transferred (shaft power) from the expander to the power-consuming compressor. The expander is a reliable, compact and highly efficient machine, which can easily be scaled to cover a wide range of process conditions.

A radial inflow turbine includes two steps of expansion, one stationary and one rotating. At peak efficiency, the stationary stage is responsible for about half the enthalpy drop and the remainder across the rotating stage. The stationary stage comprises converging nozzles mounted on an annular ring, known as inlet guide vanes or IGV's. The adjustable nozzles control the flow and respectively the cold production. The nozzles convert half of the enthalpy drop to kinetic energy in a near–isentropic expansion. The nozzles direct this high velocity flow into the rotating stage, called the impeller or wheel. The high velocity medium enters the wheel with relatively low angle of incidence, because the tip speed of the impeller approximately matches the velocity of the introduced medium. The blades extract energy when the medium enters the wheel in a radial direction and exits in an axial direction. The remaining energy can be recovered with a conical diffusor, contributing to cold production. The compressor stage should be

[14] Klaus Reuter, Integrated Machinery Systems for Cryogenic Processes Consisting of Turboexpander, Compressor, High-Frequency Motor and Generator with Magnetic Bearings

designed to balance the power produced by the expander, acting as a brake. The absorbed power determines the operating speed of the turboexpander-compressor.

Figure 5 Turboexpander/Compressor (2 TC 400/90)

A single shaft connects the expander and the compressor wheel, supported by oil lubricated radial and axial bearings. A seal element provides a barrier between the process side and the oil bearing. Use of a dry gas seal design significantly reduces the leakage of the medium, if operation is in a closed loop.

The specific design of the LIMEN™ process gives almost the same cold production in each of the warm and cold expanders. Similar speed and cold production of each expander allows operation the compressors in parallel, optimizing the performance.

The stress in the wheel is governed by the transmission torque, centrifugal force at trip speed and the axial thrust. Finite element analysis (FEA) calculations are used to optimize the stress in the wheel and determine the maximum cold production of the expander.

Recycle compressor and gas turbine driver (BHGE)

The worldwide largest BHGE LM9000 aero-derivative gas turbine, the evolution of the highly successful GE90-115B turbofan used for Boeing™ 777 commercial aircraft, is chosen to maximize the single FLNG train production with the LIMEN™ process.

The LM9000 with 43% single cycle efficiency, delivers 67 MW ISO shaft power over a wide speed range and gives the flexibility to run the FLNG plant under any condition. The capacity to generate very high torque from start-up to full speed is also a key feature for LNG mechanical drive industry providing the capabilities to start from settle-out pressure without venting the refrigerant.

	Parameter	Target
ISO performance	Power	67 MW
	Simple cycle efficiency	43%
	Operating speed	2400-3780 RPM
Emission	NOx	15 ppm
	CO	25 ppm
Maintenance	Hot gas section	36,000 hours
	Major overhaul	72,000 hours

Table 2 LM9000 performance summary table

The LM9000 has been designed for on-condition maintenance with boroscope ports for smart inspection while appropriate alloys selection throughout the engine allows longer running periods between maintenance actions. At the targeted temperatures for the combustor and turbine sections, the hot gas section exchange will be performed after 36,000 hours and the major overhaul after 72,000 hours.

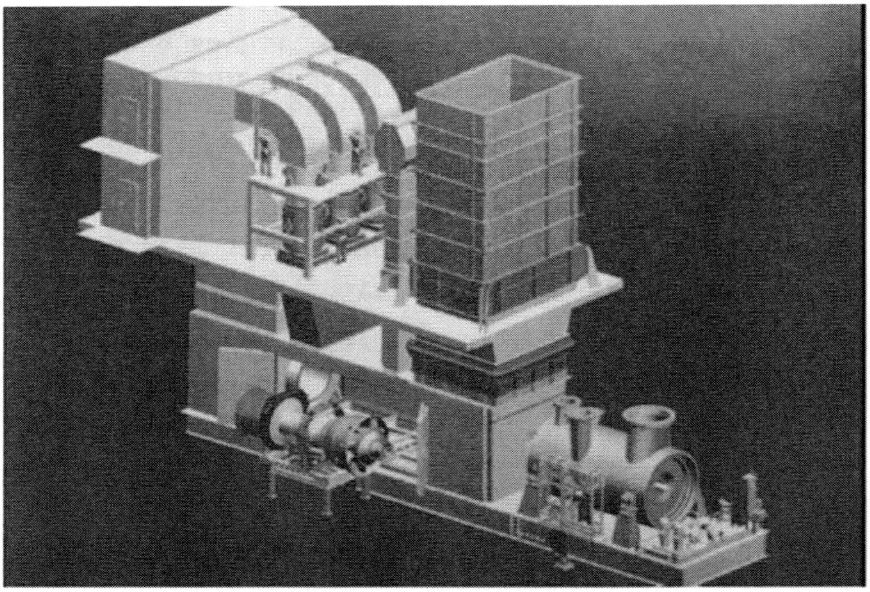

Figure 6 LM9000 centrifugal compressor outline

The LM9000 package uses the proven LM6000 SeaSmart mini-skid concept to aid in maximum "safety in design" and ergonomics for technicians to provide a controlled engine removal and installation when the vessel is in motion. Either the super-core or entire gas turbine engine can be swapped in less than 24 hours to meet high availability requirements of LNG plants. The complete package especially designed for marine application, meets stringent space requirement of FLNG layout with compact package footprint, simple balance of plant and light weight. The LM9000 gas turbine available site power at 7°C ambient temperature of ~70 MW has been saturated to maximize the LNG production.

The recycle centrifugal compressor is optimized to match the process design point with state-of-the-art BHGE compressor efficiency using a two-stages single barrel compressor casing with intermediate external process gas cooler. The selected 2BCL1406 centrifugal compressor has two stages, with six three-dimensional impellers manufactured from single forging without welding back-to-back arranged and equipped with tandem dry gas seals. Top nozzles are envisaged. The compressor is directly connected to the gas turbine driver with flexible coupling and mounted on a single baseplate. Mineral lube oil system is integrated in the baseplate to allow single lift skid.

By removing the end cover it is possible to extract the rotor diaphragm bundle assembly and to gain access to the internal components without dismounting the outer casing which remains connected to the plant piping; seals and shaft ends can be inspected and removed without opening the compressor casing. Prior to shipment, the full speed/full load string tests is performed by BHGE at its testing facility in Massa (Italy).

This compression system solution is especially favorable as contribution to the outstanding FLNG capabilities of LIMEN™ process. All major components of the envisaged compressor concept have previously been supplied by BHGE on past LNG and FLNG projects.

The same compression train concept can be used for scaled-down LIMEN™ FLNG plant throughput using smaller gas turbines models like LM6000 PF+ or LM2500/PGT25+G4 gas turbine and lower centrifugal compressor sizes (see Table 1).

Conclusions

The novel LIMEN™ process can be used to design FLNG processes with a total capacity of up to 2.5 mtpa, if two trains with 1.25 mtpa each are implemented. Thus, the equipment count and plot requirement can be significantly reduced compared to competing technologies. The thermodynamic efficiency is on the same level as a well designed N2 expander process. A major contribution to this convincing concept is rotating equipment from Cryostar (e.g. TC600/130) and BHGE (e.g. LM9000 combined with 2BCL1406).

512995 Catalytic Option(s) for Hydrotreaters Struggling with Cracked-Rich Feeds

Abstract:

To help refiners satisfy today's fuel quality specifications while hydroprocessing cracked-rich heavy distillates, development of step-out next generation hydroprocessing catalysts is needed. The application of such catalysts to the hydroprocessing of cracked-rich, heavy distillate blends should help to achieve the desired product qualities, such as diesel cetane, specific gravity, and/or product yields. Additionally, placement of the catalyst bed containing the new catalyst relative to conventional, supported hydrotreating catalysts should likewise be applicable to identify more advantageous options for the relief of processing constraints. This presentation will include reviews of these examples from pilot plant studies, highlighting the apparent advantages of at least one step-out, next generation hydroprocessing catalyst. This presentation will also cover at least some of the limitations of such a catalyst.

513669 Smart Equipment Integration for Improved Process Yield

Abstract:

A process design is presented where integration of key equipment within a hydrogen production unit produces increased process benefit. A utility consuming process in turn becomes a utility producer. Through this same flow sheet revision a problematic gas stream is re-routed to equipment that is better suited for its usage, emissions are potentially reduced, and process yield is increased.

Innovative ZoneFlow® Reactor Technology-
Gearing for Future in Steam Reforming for Hydrogen Generation

Sanjiv Ratan, ZoneFlow Reactor Technologies LLC, USA
Prof. Juray De Wilde, Universite Catholique de Louvain, Belgium

Introduction

The predominant technology for hydrogen generation over the past decades has been steam reforming of light hydrocarbons, mainly natural gas commonly called steam-methane reforming (SMR). owing to its favorable attributes This mature technology, however, has seen only incremental improvements in its catalyst design and its performance in conjunction with superior tube metallurgy and reformer design with the aim for better efficiency and reliability.

ZoneFlow™ Reactor Technologies LLC (ZFRT) has developed proprietary, radically innovative structured catalyst solutions for steam reforming, pre-reforming and post-reforming. ZoneFlow (ZF) reactors provide higher heat transfer, lower pressure drop, higher geometric surface area (GSA), more uniform flow and heat distribution, and long-term structural integrity compared to the current state-of-the-art pellet catalyst. The use of ZF reactors enables significant capital and operating cost reduction in steam reforming both for new plants as well as revamps.

Current (best available) technology

The conventional "pellet" steam reforming catalysts suffer from few inherent limitations and/or operational deficiencies. These mainly relate to a) uneven random packing voidage leading to undesired flow and temperature mal-distribution, b) catalyst attrition and breakage from thermal cycling leading to increasing pressure drop and related capacity limitations and c) pore-diffusion limited reaction path leading to curtailed intrinsic activity and lower resistance to carbon formation.

With the steam reformer being the heart of a hydrogen plant carrying out the complex steam reforming reaction involving heat, mass and momentum transfer, its reforming catalyst plays a critically important role in its thermal design, performance, reliability and cost-effectiveness.

Past attempts to develop structured steam reforming catalysts to overcome some of these limitations have suffered from the innate problems around differential expansion gaps leading to feed by-passing and disturbed boundary layer endotherm, as well as from the reaction hydraulics being away from the tube wall, lacking required heat input, thus leading to insufficient conversion and hotter tubes.

Innovative Breakthrough

ZF Reactor technology not only overcomes the above-mentioned deficiencies but also offers an unmatchable combination of higher heat transfer with substantially lower pressure drop. The ZF Reactor's unique structure also imparts a high degree of proximal flexibility which eliminates the differential thermal expansion gaps, while carrying high (metal substrate) physical strength against breakage over its lifetime.

Its nested modules provide much more geometric surface than pellets, and take away the diffusion-limited access to the layered active sites thus resulting in much higher intrinsic activity and increased resistance to carbon formation compared to pellets.

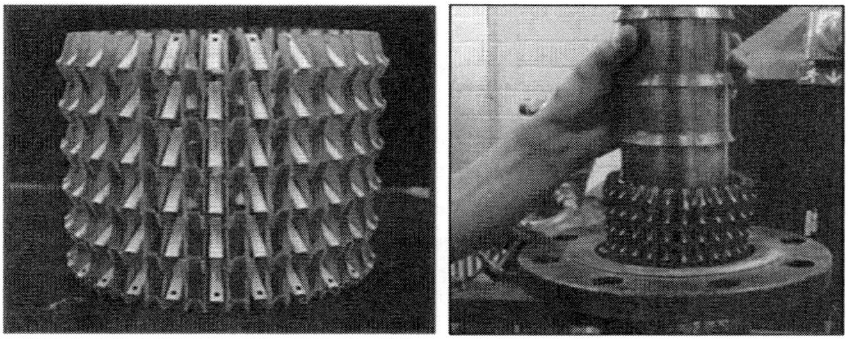

Figure-1 : ZF Reactor's typical casing and assembly

Comparative Merits of ZF Reactors vs. Conventional 'Pellet' Catalyst

ZF Reactors are advanced engineered catalyst systems based on unique 'annular' foil casing structure (see Figure-1) offering two primary advantages over conventional 'pellet' steam reforming catalysts :

- Lower pressure drop (dP) – the ZF uses dP mainly to enhance the turbulence near the tube wall, avoiding to large extent the “unutilized" dP in overcoming flow and diffusion resistance typical of random pellet packed column.
- Superior heat transfer – ZF annular structure directs the gas flow to impinge the hot surface of the tube wall which significantly enhances heat transfer and also lowers the tube skin temperatures, thus also permitting process intensification

Additionally, ZF reactors offer following significant advantages compared to pellets for revamping as well as new steam reformers.

- Much higher catalyst effectiveness for driving the endothermic steam reforming
- Uniform heat transfer and fluid flow along the tube length as well as across the tube rows; also eliminating hot spots that can restrict full-load operation and risk shorter tube life
- Long-term structural integrity, including mechanical features to accommodate thermal cycling, thus avoiding pressure drop build-up over time related to progressive deterioration of pellets

ZF development and validation milestones

Various Computational Fluid Dynamics (CFD) and Finite Element Analysis (FEA) studies have not only confirmed these properties but also allowed gaining deeper insights into the ZF design, also for optimizing the balance between these key performance parameters. ZF Reactors can be tailored to application-specific optimum combination of higher heat transfer coefficient and lower dP compared to conventional pellet catalysts as shown in Figure-2

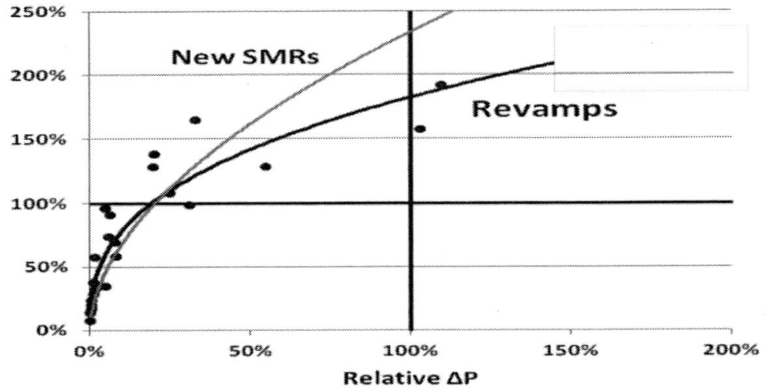

Figure-2 : Optimization of HTC against Pressure drop

ZFRT has achieved various development and validation milestones over recent years:

- Bench-scale validation at the University of California at Davis HyPaul labs.
- Independent CFD kinetic modeling for improvements and/or optimization of simultaneous heat and mass transfer against pressure drop. (Figure-3)
- FEA to establish mechanical robustness and long-term durability (Figure-4)

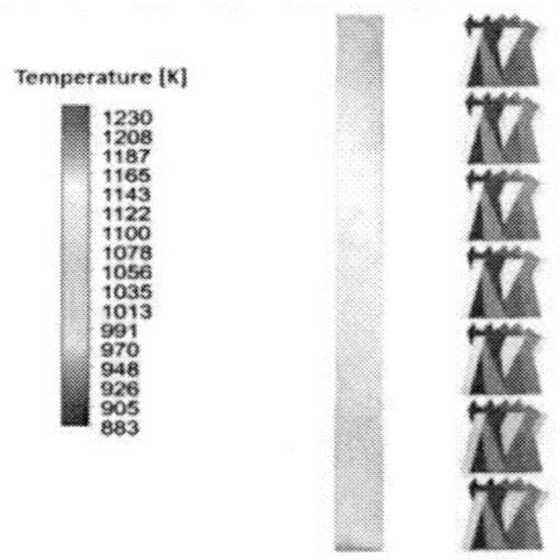

Figure-3: CFD modeling

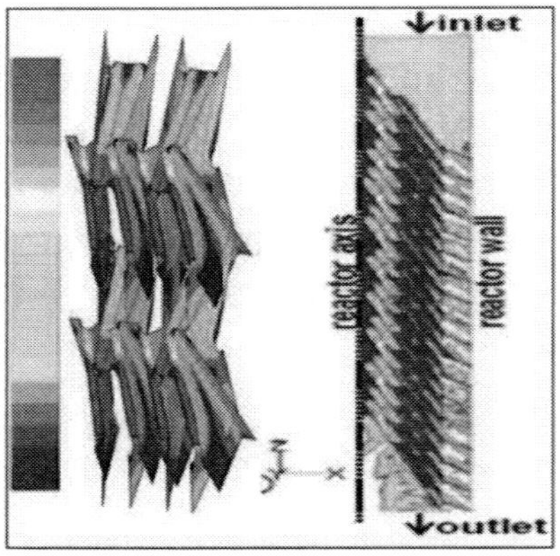

Figure-4: ZF FE Analysis Results

- Demonstration in a large commercial SMR with ZF put in 2 tubes for close to 2 years, allowing direct comparative assessment conventional BAT pellet catalyst under the same operating conditions, apart from ZF loading / unloading verification.
- Nearing completion of its own designed and built pilot plant installed within the Université Catholique de Louvain, Belgium (UCL) to further validate the performance and physical integrity of the ZF Reactor designs under wide range of conditions.

- ZF's performance parameters have been validated through advanced CFD analysis, followed by the upcoming pilot plant testing under various commercial operating conditions. ZF designs have been further tested in a commercial SMR (in 2 tubes out of more than 200) validating near 2-fold higher heat transfer together with significantly lower pressure drop compared to current pellet steam reforming catalysts.

- The ZF Reactor designs can be optimized and customized for required balance between heat transfer and pressure drop for the specific application, and can also be made to fit any commercial SMR tube size.

Added-Value ZF Solutions for Hydrogen SMRs

Following case analysis are presented covering various ZF applications :

- **ZF-Single pass (SP)**
  - De-stressing and Debottlenecking SMRs
  - Higher-flux, cost-effective and more reliable new SMRs
- **ZF-Convective pre-reforming (CPR)**
  - Unmatched in-situ retrofit for additional capacity in existing SMRs without major modifications
  - Efficient and cost-effective applications in new SMRs
- **ZF-Bayonet (B)**
  - Uniquely applicable for recuperative reforming, overcoming current barriers

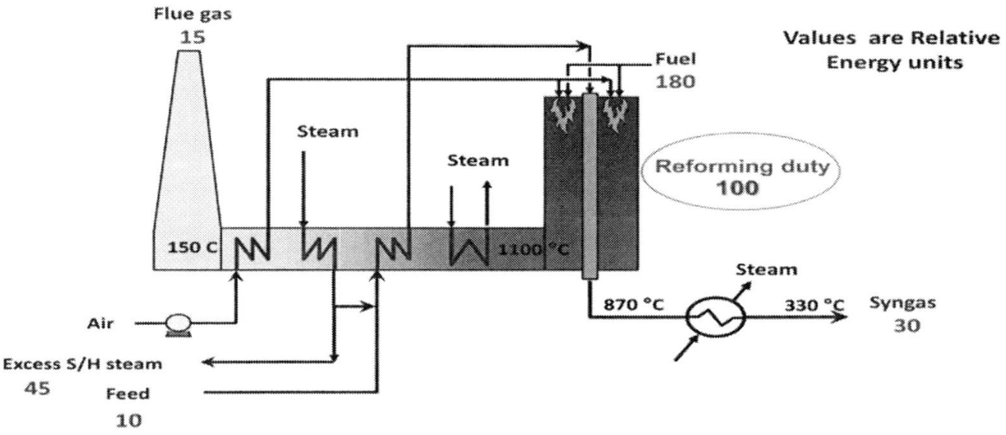

Figure-5 : Steam-methane Reformer (SMR) : Heart of a Hydrogen Plant

The common underlying advantages from ZF's lower dP and higher heat transfer coefficient include:

- higher outlet temp without increasing max. tube skin temperature (TSM)
- higher heat flux and/or higher outlet temp without increasing bridge-wall temperature, or related SMR firing duty
- lower Approach to Equilibrium

Case 1 Stressed-SMR attributes and make-over using ZF-SP reactors

- Pressure drop (or its build up) limiting throughput to under-nameplate capacities
- Loss of catalyst activity → hotter tubes limiting outlet temperature
- Catalyst attrition / settling from thermal cycling
- Carbon formation / hot spots
- Impact on (remaining) tube life
- Constrained reforming capacity
- Replacing (pellet) catalyst in these SMRs with ZF-SP Reactors can overcome these deficiencies, as summarized in following case analysis :

Table-1: SMR De-Stressing	SMR Design	Stressed Operation	ZF-SP replacing pellets
Relative Capacity, %	**100**	95	100
Capacity limitations	-	dP, TSM	removed
S/C Ratio	3.0	3.3	2.8
Outlet temp, C	860	843	868
Approach to Equilibrium EOR C	-10	-12	-7
CH4 slip, dry vol %	5	5	5
Radiant Pressure drop, bar	2.8	2.8	2.3
Relative Radiant duty %	100	97	99
Avg heat flux kW/m2	78	74	78
Bridgwall temp, C	1000	980	998
TSM (avg) C	935	935	926

Case 2 ZF-SP Reactors for SMR Debottlenecking

- Achieve capacity in excess of nameplate (~ 105 %), while utilizing design margins in SMR burners and fans
- Higher average heat flux without exceeding tube design temperature
- Improved temperature uniformity
- Extended tube life and improved reliability
- Better catalyst performance and "life cycle" costs
- Optimized operation and enhanced reliability

Table-2 : SMR Debottlenecking	Reference	ZF-Radiant
Relative Capacity, %	100	105
Capacity limitations	dP, TSM	removed
S/C Ratio	3.0	2.8
Outlet temp, C	865	870
Approach to Equilibrium C	-10	-7
CH4 slip, vol %	5.5	5.7
Radiant Pressure drop, bar	2.8	2.5
Relative Radiant duty %	100	104
Avg heat flux kW/m2	75	78
Bridgwall temp, C	**1008**	**1004**
TSM C	**940**	**938**

Case 3 ZoneFlow™ Convective Pre-reforming (ZF-CPR)

ZFRT has also designed and engineered an ultra-low pressure drop structured catalytic reactor with high surface area / low cross section for non-adiabatic low-temperature convective pre-reforming (CPR).

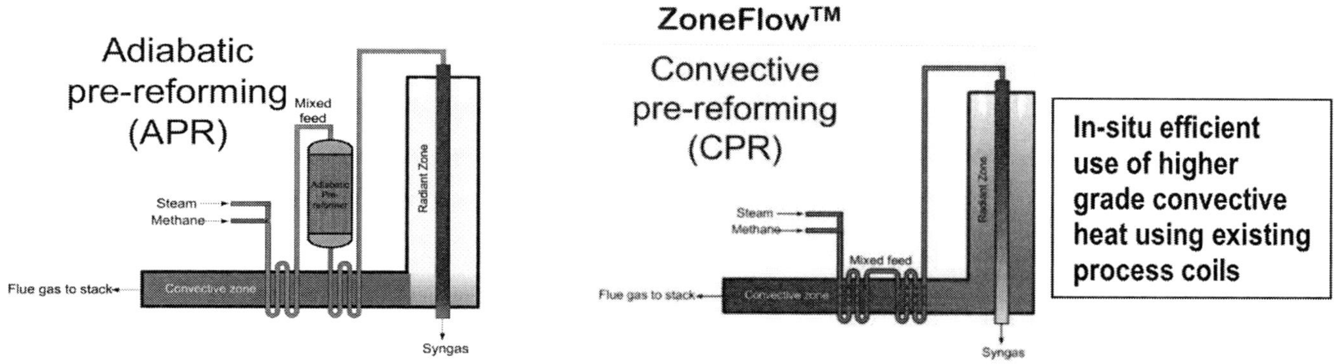

- In-situ horizontal loading in the mixed feed preheat (MFPH) coil in the convection
- Optimization of pressure drop in combination with ZF-SP in SMR radiant tubes
- For revamps, 8-12% additional reforming without increasing SMR firing duty
- For New SMRs, 10-15% smaller SMR and proportionately also the steam system.
- No concerns over horizontal loading catalyst settling, attrition and related feed by-passing
- Shifts steam loads to Utilities and Offsites , thus also lowering core-plant C-footprint

Table 3: ZF-CPR Integration		Existing	ZF + CPR
Relative Capacity,	**%**	**100**	**115**
S/C Ratio		**3.0**	**2.8**
SMR inlet temp,	C	550	550 / 575 *
SMR outlet temp,	C	870	870 / **878** *
Approach to Equilibrium EOR	C	-10	-7
CH4 slip,	dry vol %	4.8	5.4 / 5.0
Radiant Pressure drop,	**bar**	**2.5**	**2.3**
Avg heat flux	kW/m2	80	80 - 82
Relative Radiant duty	**%**	**100**	**100 / 103**
Bridgewall temp,	**C**	**1020**	**1009 / 1018**
TSM	**C**	**950**	**938 / 948**

ZF-Bayonet (B) for Recuperative Reforming

ZF Reactors are inherently suited for "bayonet" (tube-in-tube) recuperative reforming based on their annular casing design. The reformed process gas flows out through the center (core) tube providing part of the high grade heat recovery for steam reforming , thus lowering the fired duty as well as size of the reformer by 15-20% . ZF also overcomes the main risk of current "pellet" catalyst crushing from differential thermal expansion as an unmatched technology based on its robust metallic substrate and flexible casing structure.

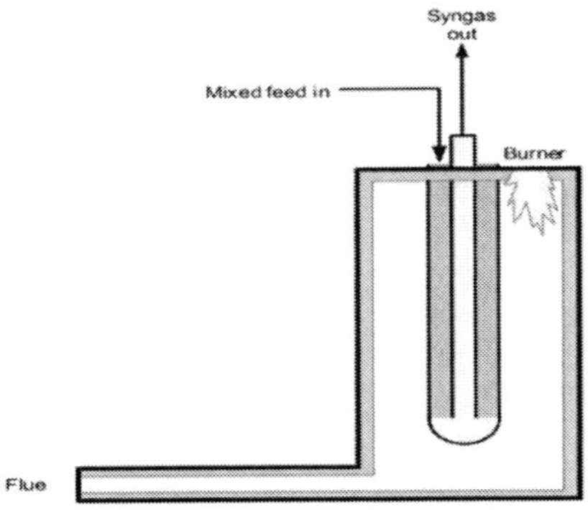

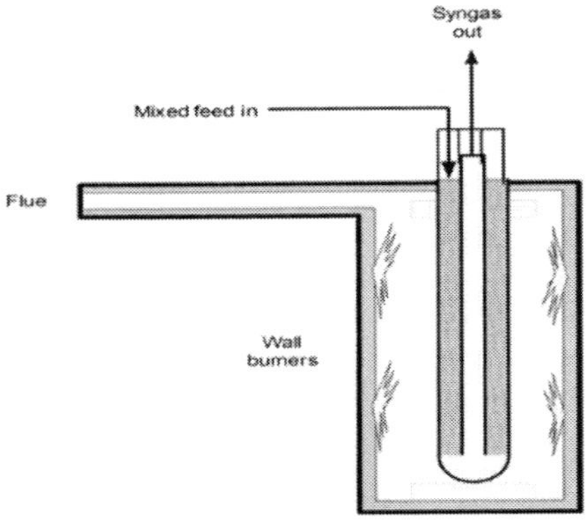

Economic Benefits and Applications

These ZF Reactor merits can translate into attractive benefits ranging from relieving "stressed" reformers, extending reformer tube life and capacity revamps to lowering SMR related OPEX and CAPEX. It covers the capacity span from large hydrogen generation plants for the refining sector (whether new or revamp of existing plants) to the advanced solutions for small scale distributed hydrogen plants for fuel cells and futuristic hydrogen energy landscape. It's advanced merits also cover the related Syngas production for methanol and GTL plants, including mini modularized units for remote gas monetization as well as Flare gas reduction.

Based on the energy costs and economic investment / IRR criteria for new plants and the level of modifications in revamping of existing plants, each application needs to be evaluated on a case-to-case basis. in terms of Opex and Capex analysis to select the most appropriate, conforming and economically attractive (ZF) solution. Other important case-specific considerations may include stand-alone units with no export steam, lowering of carbon footprint, SMR intensification, long term feed & product pricing, revamp project downtime as well as environmental aspects for overall optimization of hydrogen generation.

Conclusions

ZoneFlow™ Reactor Technology (ZFRT) developed innovative, and differentiating structured catalyst technologies with demonstrated improvement in heat transfer, pressure drop and catalyst effectiveness. Further , the new pilot plant designed and constructed at UCL, Belgium and also the collaborative efforts with some of the prominent players in the syngas field, will allow further rigorous testing of ZF reactors under commercial conditions for application in steam methane reforming. The portfolio also includes pre-reforming and recuperative reforming. ZF reactor technologies offer attractive applications and promising solutions in steam reforming and for realizing advanced options, thus facilitating cost-effective , efficient, reliable and distinctly attractive hydrogen generation.

513253 Catalyst Characterization in the Molecular-Level Kinetic Modeling of Hydroprocessing Reactors

Abstract:

The commercial upgrading of conventional and unconventional hydroprocessing feeds to valuable petrochemicals and fuels usually involves a series of reactors, each of which can contain multiple beds with several different layers of catalysts. A multiphase hydroprocessing reactor system with molecular detail of the feed and products was modeled using an in-house software, the Kinetic Modeler's Toolbox. A reaction network of hydroprocessing feeds with $O(10^3)$ species and $O(10^4)$ reactions that includes the detailed hydroprocessing chemistry (saturation, hydrocracking, ring opening, hydrodesulfurization, hydrodenitrogenation, and hydrodeoxygenation) was generated. Several coke formation pathways and metal deposition reactions were also modeled to account for time-on-stream catalyst site reduction. The initial flow rates of the feed were calculated by defining an attribute, or structural moiety, probability density function tree and optimizing the difference between the feed experimental and simulated properties. A set of catalytic rate law based material balances, one for each species, and an energy balance along with the feed flow rates was used to define the initial value problem to be solved in the kinetic model. Each reactor bed was modeled as a series of plug-flow pseudo-reactors with side-by-side reaction and vapor-liquid equilibrium, where each pseudo-reactor represents a separate catalyst layer. The mixing of the hydrogen quench streams between beds was modeled with a mass balance and an enthalpy balance. Additionally, a complementary selectivity-activity modifier model was used to model the catalyst deactivation on the longer timescale of the catalyst life. Linear free-energy relationships (LFERs) and quantitative structure-property relationships (QSPRs) were used to reduce and tune the parameters of the LHHW rate equations. The LFER concept was extended to catalyst families in order to further minimize the number of parameters and data needed for different catalysts[1]. The final model represents a real refinery reactor simulation over a range of process conditions like space velocity, temperature, pressure, and hydrogen flow. It also allows for scenario studies to test catalyst deactivation for different feed compositions and the impact of changing catalyst layer order, bed height, and quench rates.

[1] Horton, S.R., Klein, M.T. "Reaction and Catalyst Families in the Modeling of Coal and Biomass Hydroprocessing Kinetics." *Energy & Fuels*, 28:1 (2014), 37–40.

512390 Molecular Kinetic Modeling in Hydrocracking

Abstract:

Hydrocracking is in an exceedingly competitive commercial environment which necessitates licensors to provide increasingly sophisticated technical solutions. In recent years, hydrocracking solutions have expanded to catalyst stacking options, configuration studies and flexible product objectives. Next generation discoveries will depend on the ability to understand how different molecules interact with different catalysts and materials.

Historic limitations in analytical capabilities have driven most to implement lumped kinetic models where the molecules are grouped by their physical properties such as boiling point. While this approach enables accurate estimates of product yields within the scope of experimental data used to generate the model, it inevitably obscures molecular information due to multicomponent nature of each lump rendering the model without predictive capability for variable feeds, process configurations and materials.

While molecular kinetic models offer the potential to be predictive, they are immensely difficult to construct due to the vast number of species, reactions and associated rate constants. Recent advancements in analytical capabilities to enhance feed and product characterization as well as developments of mathematical methods to reduce the number of rate constants have made it possible to attempt feedstock-independent models that care applicable in around the commercially relevant feed and process conditions.

In this talk, hydrocracking molecular kinetic modeling and itâ€™s applications will be presented. Case studies will be shown to demonstrate the capabilities of the model in development of novel technical solutions.

513618 Lumped Kinetic Model for Hydrocracking of Heavy Oils over Oil-Dispersed Mo-Based Catalyst

Abstract:

A 5-lump (unconverted residue, vacuum gas oil, distillates, naphtha and gases) kinetic model for hydrocracking of heavy oils is developed. To obtain the experimental data, a continuous-stirred tank reactor has been used and the experiments have been carried out using Mo-based oil-dispersed catalyst precursor and vacuum residue from the commercial refinery in South Korea. The reaction temperature has varied in the range of 425 â€" 440 Â°C and the reactor pressure has varied in the range of 140 â€" 160 bar, which are the ordinary operating window of the slurry hydrocrackers. For the efficient parameter estimation, the efficient optimization algorithm combining the genetic algorithm and the deterministic algorithm have been adopted due to the nonlinear and high dimensional object function for the parameter estimation. The prediction results by the developed kinetic model has shown the good agreement with the experimental values so this work is considered to contribute to designing an appropriate slurry hydrocracker.

512904 Change of Product Yield and Quality with a Variation of Residue Conversion in a Slurry-Phase Hydrocracking of Vacuum Residue

Abstract:

Slurry-phase hydrocracking (SPH) technology is one of the promising technologies for upgrading of heavy oils in the refining area. Since it uses a nano-sized dispersed catalysts the asphaltene, which is known for the difficult component to be treated by heterogeneous catalysts, can be hydrogenated and converted into a light molecular sized component. However, SPH is still on-going demonstration test in a commercial field and it is not found from the literature what characteristics SPH has in a once-through operation mode. In this respects, this study investigated the change of product yield and quality with a variation of residue conversion through a once-through mode in a slurry-phase hydrocracking of vacuum residue.

The experiment was carried out with a function of reaction temperature from 420 to 440 â„ƒ and liquid hourly space velocity (LHSV, 0.3 â€“ 0.5 1/h), hydrogen partial pressure (80-160 bar), H_2 to feed ratio (600-2000 Nm^3/kl) with molybdenum catalyst concentration of 500 wt.ppm based on metal content on feed in a 4L slurry bubble column reactor. For the analysis of the product yield and properties, gas composition was determined using a gas chromatograph (Agilent 7890) while the liquid properties for the API (D 5002-99), sulfur and metals (nickel and vanadium, D4294-03), micro-carbon residue (MCR, D4530-06), asphaltene (normal heptane insoluble content, D2007), and boiling point distribution (D7500) were all analyzed according to standard ASTM methods. Additional, the amount of sediments and the remained catalyst concentration in the liquid product was examined after reaction.

512609 Dynamic Modeling of the Hybrid LNG Terminal Coupling NGL Recovery and LNG Re-Gasification

Abstract:

With the high demands of shale gas and promising development of LNG terminals, a lot of research has focused towards the process development for effective recovery of C2 and heavier hydrocarbons (NGL). Shale gas requires a large amount of cold energy to cool down and recover the NGL; and the LNG re-gasification process requires a lot of heat energy to evaporate for NGL recovery. Thus, coupling the shale gas NGL recovery process and LNG re-gasification process, for utilizing the cold energy from LNG re-gasification process to assist NGL recovery from shale gas has significant economic benefits on both energy saving and high value product recovery. Wang et al. developed new conceptual design of such coupled process in 2013 [1] and later developed an improved optimal design considering uncertainties in 2014 [2]. In this paper, the virtual validation of this novel conceptual design will be performed via plant-wide dynamic simulations to examine its feasibility and optimality. First of all, a steady-state model of this new process has been developed based on the plant design proposed by Wang et al. with Aspen plus®. The process couples the NGL recovery and LNG re-gasification processes with efficient heat integrations. Next, the dynamic model of the entire process has been developed and tested using Aspen Plus Dynamics® with newly developed control strategies. After that, the flexibility of the dynamic plant model is examined and validated under disturbances of ±25% changes of the shale gas feed flowrate. The energy consumptions associated with different operating conditions will be also investigated. This study successfully demonstrated a novel conceptual design of the hybrid LNG terminal with the integration of both NGL recovery and LNG re-gasification processes.

References

1. Wang, M.; Zhang, J.; Xu, Q. A Novel Conceptual Design by Integrating NGL Recovery and LNG Re-gasification Processes for Maximum Energy Savings. *AIChE Journal*. **2013**, 59, 4673-4685.
2. Wang, M.; Xu, Q. Optimal Design and Operation for Simultaneous Shale Gas NGL Recovery and LNG Re-gasification under Uncertainties. *Chemical Engineering Science*. **2014**, 12, 130-142.

512818 Remove/ Use Gigatons of CO2 from Flue Gas withoutÂ CO2 Recovery or Purification

Abstract:

No, its not a perpetual motion machine. This paper will introduce a watershed patent pending process where raw flue gas, when routed and applied properly in a sustained fashion will triple the rate of production of an ubiquitous sustainable "green" commodity product easily absorbed by the market. Beneficially CO2 in the flue gas is reduced with very minimal energy input, yielding a net negative GHG contribution most cost effectively, as it does not require CO2 recovery or purification. To be successful in reducing Gigatons of CO2 per year levels, it will require significant investment by industry (and government) in infrastructure, as expected for any solution of the growing magnitude of the GHG dilemma. In effect this is a solution with an infrastructure impact just like that required by the culprit utilities of electricity and hydrocarbon fuel before their use became widespread.

512597 Thermodynamics Analysis Based Optimal Design and Operation of Dmr Refrigeration System for Natural Gas Liquefaction Process

Abstract:

Thermodynamics Analysis Based Optimal Design and Operation of DMR Refrigeration System for Natural Gas Liquefaction Process

Mozammel Mazumder, Qiang Xu*, and Srinivas Palanki

Dan F. Smith Department of Chemical Engineering

Lamar University, Beaumont, Texas 77710, USA

Abstract

The liquefied natural gas (LNG) supply chain has diversified the global gas market previously dominated by pipeline gas suppliers and improved the security of energy supply of many consuming nations. The traditional LNG processes mainly include cascade (Jensen, et al.; 2006), nitrogen expansion cycle (Remeljej, et al., 2006) and mixed refrigeration cycle (MRC) (Wang, et al., 2007). In this paper, a new methodology for LNG liquefaction process introduced combining turboexpansion, dual mixed refrigeration (DMR), Joule-Thompson expansion targeting energy consumption minimization. It also described thermodynamic analysis based study of the minimization of the energy consumption of the LNG process. The developed methodology contains three major tasks: (i) modeling and simulation of an existing LNG case (base case); (ii) thermodynamics and mathematical analysis for solution identification; and (iii) modeling and optimization of the newly developed LNG process. In the first stage, based on the modeling and simulation of the base case, the operating status of the refrigerant and natural gas streams of the entire NG liquefaction process are obtained. Aspen HYSIS version 8.8 with PENG-ROB equation of state is employed for the modeling. In the second stage, the energy consumption roadmap is explored through thermodynamic analysis where the temperature and specific enthalpy (T-H) diagram is employed. Based on thermodynamic studies, the energy saving opportunities to improve the base case (i.e., new LNG process) has been identified. In the third stage, a rigorous optimization model is developed to obtain the optimal solution of the new LNG process. Certainly, the optimization results will be validated through rigorous process simulation and thermodynamic analysis. Finally, the energy analysis and economic evaluation between the base and the optimal case will be performed.

References

Jansen, J. B.; Skogestad, S. Optimal operation of a mixed fluid cascade LNG plantâ€. *Comp. Aid. Chem. Eng.* **2006**, *21*, 1569-1574.

Remeljej, C.W.; Hoadley, A. F. A. An exergy analysis of small-scale liquefied natural gas (LNG)

liquefaction processes. *Energy*. **2006**, *31*, 2005-2019.

Wang, K.; Xu, F. Y.; Li, H. Y. Dynamic performance simulation of plate-fin heat exchangers in small scale MRC-LNG plant. *Cryogenics*. **2007**, *3*, 44-49.

512434 Reboilers with Plate-Fin Brazed Aluminum Heat Exchangers: Best Practices and Considerations for Design and Operation

Abstract:

Brazed Aluminum Heat Exchangers (BAHX) are used in cryogenic reboiler applications for natural gas processing and ethylene production industries. In these applications, plant operating conditions are changing constantly due to changing economic factors or inlet gas composition. This makes operation of the equipment challenging.

This presentation will outline best practices and considerations for the design and operation of BAHX reboilers that will help maximize the lifetime of the BAHX and help promote more robust plant operation. Considerations will include piping, flow stability, and heat exchanger design. A summary and references for operating guidelines will be included. Some examples of consequences from not following best practices will be shown.

512311 Shale Gas Techno-Economic Analysis: Designing Separation Units to Handle Feedstock Variability

Abstract:

Shale gas is abundant in the U.S. and this presents many opportunities in the chemical, petrochemical, and fuel industries. As with conventional natural gas, shale gas is primarily methane, but also may contain significant compositions of higher molecular weight hydrocarbons, and other impurities. This variability presents a technical challenge to processing the gas. There is a tradeoff between the size (and therefore cost) of processing units and the ability to handle a wider range of compositions. This research aims to find a design to optimize profitability based on processing differing raw shale gas compositions.

In this work a literature review of conventional and alternate technologies for processing shale gas was performed. Composition of the raw shale gas was based on literature data for wells in the Barnett Shale play located northwest of Dallas. Process simulators such as ProMax, ASPEN HYSYS, and ASPEN Plus were used to construct base-case scenarios, and to carry out process simulations for differing compositions with any necessary process modifications. A techno-economic analysis was done to estimate the associated costs and profitability of all the scenarios. Once this was done an optimal design was chosen and validated for a variety of compositions. Additionally a sensitivity analysis was done to see the effect of the various costs (fixed costs, variable costs, raw shale gas price, product prices) on profitability.

AUTHOR INDEX

AIChE
120 Wall Street, FL 23
New York, NY 10005-4020

CURRAN ASSOCIATES INC.
proceedings.com

ISBN 978-1-5108-6426-9